Communications
in Computer and Information Science 1740

More information about this series at https://link.springer.com/bookseries/7899

Katerina Zdravkova · Lasko Basnarkov (Eds.)

ICT Innovations 2022

Reshaping the Future Towards a New Normal

14th International Conference, ICT Innovations 2022
Skopje, Macedonia, September 29 – October 1, 2022
Proceedings

Springer

Editors
Katerina Zdravkova ⓘD
Saints Cyril and Methodius University
of Skopje
Skopje, North Macedonia

Lasko Basnarkov ⓘD
Saints Cyril and Methodius University
of Skopje
Skopje, North Macedonia

ISSN 1865-0929 ISSN 1865-0937 (electronic)
Communications in Computer and Information Science
ISBN 978-3-031-22791-2 ISBN 978-3-031-22792-9 (eBook)
https://doi.org/10.1007/978-3-031-22792-9

This Springer imprint is published by the registered company Springer Nature Switzerland AG
The registered company address is: Gewerbestrasse 11, 6330 Cham, Switzerland

Preface

This volume contains the best papers and the extended abstracts of keynote talks of 14th ICT Innovations 2022, which was held at the Faculty of Computer Science and Engineering (FCSE) in Skopje from September 29 to October 1, 2022. After two years of online events due to the restrictions caused by COVID-19 pandemic, this edition was hybrid, uniting more than 50 researchers in-person and additional 50 participants who joined the conference, the poster session and the satellite workshops virtually. We cordially thank FCSE computing center staff, who created and maintained an impeccable working environment.

The focal topic of 14th ICT Innovations 2022 was "Reshaping the future towards a new normal". Four keynote speakers prepared talks in line with the topic.

By the final deadline on 30 June 2022, 42 papers prepared by 119 authors from 15 countries were submitted. They were thoroughly examined and graded by 108 reviewers from 33 countries. For each paper, four to five reviewers were assigned, all coming from different countries than the authors. The review process was double-blind, increasing the objectivity of the judgements. The best 14 papers are part of these proceedings. Additionally, 15 papers and three posters were accepted. They are published in the conference web proceedings.

The papers were related to many topic areas, which were clustered into the follow-ing: artificial intelligence applications; business and application software; computer sciences and edge computing; education; health (medical) informatics; and signal processing and machine learning.

For the first time, a best paper was awarded. A joint jury consisting of program and scientific committee members made the decision to award the paper: "An explo-ration of autism spectrum disorder classification from structural and functional MRI images".

The conference was organized by the Association for Information and Communi-cation Technologies (ICT-ACT), whose main goal is to support the development of information and communication technologies, especially in the area of education, research and application of innovative technologies. The conference was supported and hosted by the Faculty of Computer Science and Engineering of the University Ss. Cyril and Methodius in Skopje, the oldest and the best university in Macedonia. It was officially opened by Ordan Chukaliev, vice-rector for international cooperation.

We cordially thank all the authors, chairpersons, scientific and program committee members, reviewers, sub-reviewers, technical committee members, and, in particular, ICT-ACT chair Sasho Gramatikov, whose enthusiasm, proactivity and considerate support inspired us to finish our obligations timely.

We were privileged that our distinguished keynote speakers accepted the invitation and gave the impressive talks: "AI ethics as ecosystems ethics? Towards a new nor-mal of understanding ethical issues in AI" (Bernd Carsten Stahl, De Montfort Univer-sity, UK; online); "Strange roads" (Kai Kimppa, University of Turku, Finland; in-person);

"The new normal: innovative informal digital learning after the pandemic" (John Traxler, University of Wolverhampton, UK; online); and "AI-based approaches in processing big data in precision medicine" (Mirjana Ivanovic, University of Novi Sad, Serbia; in-person). We are grateful to all the presenters and conference partici-pants for their provoking contributions. They helped to made ICT Innovations 2022 a successful event.

We look forward to seeing you at the jubilee 15th edition of the conference next year.

October 2022   Katerina Zdravkova
                                                 Lasko Basnarkov

Organization

General Chairs

Katerina Zdravkova   Ss. Cyril and Methodius University in Skopje, North Macedonia

Lasko Basnarkov   Ss. Cyril and Methodius University in Skopje, North Macedonia

Scientific Committee

Ljupcho Antovski   Ss. Cyril and Methodius University in Skopje, North Macedonia

Goce Armenski   Ss. Cyril and Methodius University in Skopje, North Macedonia

Danco Davcev   Ss. Cyril and Methodius University in Skopje, North Macedonia

Dejan Gjorgjevikj   Ss. Cyril and Methodius University in Skopje, North Macedonia

Sasho Gramatikov   Ss. Cyril and Methodius University in Skopje, North Macedonia

Boro Jakimovski   Ss. Cyril and Methodius University in Skopje, North Macedonia

Program Committee

Jugoslav Achkoski   Military Academy "General Mihailo Apostolski", North Macedonia

Nevena Ackovska   Ss. Cyril and Methodius University in Skopje, North Macedonia

Marco Aiello   University of Stuttgart, Germany

Luis Alvarez Sabucedo   University of Vigo, Spain

Ljupcho Antovski   Ss. Cyril and Methodius University in Skopje, North Macedonia

Goce Armenski   Ss. Cyril and Methodius University in Skopje, North Macedonia

Hrachya Astsatryan   National Academy of Sciences of Armenia, Armenia

Amelia Badica   University of Craiova, Romania

Verica Bakeva Smiljkova   Ss. Cyril and Methodius University in Skopje, North Macedonia

Antun Balaz   Institute of Physics Belgrade, Serbia

Lasko Basnarkov   Ss. Cyril and Methodius University in Skopje, North Macedonia

Slobodan Bojanic	Universidad Politécnica de Madrid, Spain
Andrej Brodnik	University of Ljubljana, Slovenia
Francesc Burrull	Universidad Politecnica de Cartagena, Spain
Ivan Chorbev	Ss. Cyril and Methodius University in Skopje, North Macedonia
Betim Cico	EPOKA University, Albania
Emmanuel Conchon	University of Limoges, France
Fisnik Dalipi	Linnaeus University, Sweden
Robertas Damasevicius	Silesian University of Technology, Poland
Antonio De Nicola	ENEA, Italy
Aleksandra Dedinec	Ss. Cyril and Methodius University in Skopje, North Macedonia
Boris Delibašić	University of Belgrade. Serbia
Vesna Dimitrova	Ss. Cyril and Methodius University in Skopje, North Macedonia
Ivica Dimitrovski	Ss. Cyril and Methodius University in Skopje, North Macedonia
Milena Djukanovic	Univerzitet Crne Gore, Montenegro
Martin Drlik	Constantine the Philosopher University in Nitra, Slovakia
Tome Eftimov	Jozef Stefan Institute, Slovenia
Amjad Gawanmeh	University of Dubai, United Arab Emirates
Ilche Georgievski	University of Stuttgart, Germany
John Gialelis	University of Patras, Greece
Hristijan Gjoreski	Ss. Cyril and Methodius University in Skopje, North Macedonia
Dejan Gjorgjevikj	Ss. Cyril and Methodius University in Skopje, North Macedonia
Rossitza Goleva	New Bulgarian University, Bulgaria
Sasho Gramatikov	Ss. Cyril and Methodius University in Skopje, North Macedonia
Andrej Grgurić	Ericsson Nikola Tesla d.d., Croatia
David Guralnick	International E-Learning Association, USA
Marjan Gusev	Ss. Cyril and Methodius University in Skopje, North Macedonia
Yoram Haddad	Jerusalem College of Technology, Israel
Fu-Shiung Hsieh	Chaoyang University of Technology, Taiwan
Ladislav Huraj	University of SS. Cyril and Methodius in Trnava, Slovakia
Hieu Trung Huynh	Industrial University of Ho Chi Minh City, Vietnam
Sergio Ilarri	University of Zaragoza, Spain
Natasha Ilievska	Ss. Cyril and Methodius University in Skopje, North Macedonia
Mirjana Ivanovic	University of Novi Sad, Serbia
Smilka Janeska - Sarkanjac	Ss. Cyril and Methodius University in Skopje, North Macedonia

Mile Jovanov	Ss. Cyril and Methodius University in Skopje, North Macedonia
Milos Jovanovik	Ss. Cyril and Methodius University in Skopje, North Macedonia
Vacius Jusas	Kaunas University of Technology, Lithuania
Elinda Kajo Mece	Polytechnic University of Tirana, Albania
Slobodan Kalajdziski	Ss. Cyril and Methodius University in Skopje, North Macedonia
Kalinka Kaloyanova	University of Sofia, Bulgaria
Ivan Kitanovski	Ss. Cyril and Methodius University in Skopje, North Macedonia
Magdalena Kostoska	Ss. Cyril and Methodius University in Skopje, North Macedonia
Bojana Koteska	Ss. Cyril and Methodius University in Skopje, North Macedonia
Arianit Kurti	Linnaeus University, Sweden
Petre Lameski	Ss. Cyril and Methodius University in Skopje, North Macedonia
Igor Ljubicic	University of Zagreb, Croatia
Suzana Loshkovska	Ss. Cyril and Methodius University in Skopje, North Macedonia
Ana Madevska Bogdanova	Ss. Cyril and Methodius University in Skopje, North Macedonia
Gjorgji Madjarov	Ss. Cyril and Methodius University in Skopje, North Macedonia
Smile Markovski	Ss. Cyril and Methodius University in Skopje, North Macedonia
Hristina Mihajloska	Ss. Cyril and Methodius University in Skopje, North Macedonia
Aleksandra Mileva	Goce Delčev University of Štip, North Macedonia
Biljana Mileva Boshkoska	University of Ljubljana, Slovenia
Georgina Mirceva	Ss. Cyril and Methodius University in Skopje, North Macedonia
Miroslav Mirchev	Ss. Cyril and Methodius University in Skopje, North Macedonia
Kosta Mitreski	Ss. Cyril and Methodius University in Skopje, North Macedonia
Pece Mitrevski	University "St. Kliment Ohridski", Bitola, North Macedonia
Irina Mocanu	Politehnica University of Bucharest, Romania
Andreja Naumoski	Ss. Cyril and Methodius University in Skopje, North Macedonia
Novica Nosović	University of Sarajevo, Bosnia and Herzegovina
Dilip Patel	London South Bank University, UK
Matus Pleva	Technical University of Košice, Slovakia
Florin Pop	Politehnica University of Bucharest, Romania

Zaneta Popeska	Ss. Cyril and Methodius University in Skopje, North Macedonia
Aleksandra Popovska-Mitrovikj	Ss. Cyril and Methodius University in Skopje, North Macedonia
Marco Porta	University of Pavia, Italy
Ustijana Rechkoska-Shkoska	UIST—Ohrid, North Macedonia
Manjeet Rege	University of St. Thomas, USA
Panche Ribarski	Ss. Cyril and Methodius University in Skopje, North Macedonia
Blagoj Ristevski	University St Kliment Ohridski Bitola, North Macedonia
Sasko Ristov	University of Innsbruck, Austria
David Šafránek	Masaryk University, Czech Republic
Jatinderkumar Saini	Narmada College of Computer Application, India
Simona Samardjiska	Radboud University, The Netherlands
Snezana Savoska	University St Kliment Ohridski Bitola, North Macedonia
Loren Schwiebert	Wayne State University, USA
Vladimír Siládi	Matej Bel University, Slovakia
Josep Silva	Universitat Politècnica de València, Spain
Manuel Silva	Polytechnic of Porto and INESC TEC CRIIS, Portugal
Monika Simjanoska	Ss. Cyril and Methodius University in Skopje, North Macedonia
Dejan Spasov	Ss. Cyril and Methodius University in Skopje, North Macedonia
Riste Stojanov	Ss. Cyril and Methodius University in Skopje, North Macedonia
Milos Stojanovic	Visoka tehnicka skola Nis, Serbia
Stanimir Stoyanov	University of Plovdiv "Paisii Hilendarski", Bulgaria
Biljana Tojtovska	Ss. Cyril and Methodius University in Skopje, North Macedonia
Dimitar Trajanov	Ss. Cyril and Methodius University in Skopje, North Macedonia
Ljiljana Trajkovic	Simon Fraser University, Canada
Vladimir Trajkovik	Ss. Cyril and Methodius University in Skopje, North Macedonia
Denis Trcek	University of Ljubljana, Slovenia
Christophe Trefois	University of Luxembourg, Luxemnourg
Katarina Trojacanec Dineva	Ss. Cyril and Methodius University in Skopje, North Macedonia
Elena Vlahu-Gjorgievska	University of Wollongong, Australia
Shuxiang Xu	University of Tasmania, Australia
Eftim Zdravevski	Ss. Cyril and Methodius University in Skopje, North Macedonia

Katerina Zdravkova   Ss. Cyril and Methodius University in Skopje,
                           North Macedonia
Xiangyan Zeng   Fort Valley State University, USA

Technical Committee

Jovana Dobreva   Ss. Cyril and Methodius University in Skopje,
                           North Macedonia
Dimitar Kitanovski   Ss. Cyril and Methodius University in Skopje,
                           North Macedonia
Boris Mantov   Ss. Cyril and Methodius University in Skopje,
                           North Macedonia
Damjan Mishevski   Ss. Cyril and Methodius University in Skopje,
                           North Macedonia
Ana Todorovska   Ss. Cyril and Methodius University in Skopje,
                           North Macedonia

Keynote Talks
(Extended Abstracts)

AI Ethics as Ecosystems Ethics? Towards a New Normal of Understanding Ethical Issues in AI

Bernd Carsten Stahl🆔

University of Nottingham, Nottingham, UK
Bernd.Stahl@nottingham.ac.uk

Abstract. This brief overview document outlines some of the key ideas developed with regards to how we may look at questions of the ethics of AI in future. It is based on prior research undertaken in the context of the EU project SHERPA which explored empirical and conceptual aspects of the current engagement with AI. It proposes that the use of the metaphor of an ecosystem may be helpful in understanding AI. If accepted, this position raises numerous follow-on questions about how ethics of ecosystems can be conceptualized and implemented.

Keywords: Artificial intelligence · Ecosystems · Ethics

1 Introduction

When we talk about the "new normal" in the autumn of 2022, we typically refer to the way things will go in future, after the ebbing of the Covid pandemic when individuals and societies have found ways of coping with this new and persistent health threat. It typically refers to questions such as the future of work, changing travel practices or heightened attention to the importance of personal hygiene in preventing infections. The new normal can also be interpreted more broadly, looking at other social changes, e.g. the use and consumption of energy in light of the looming climate catastrophe or ways in which societies deal with fuel and other poverty.

In this short abstract, I will propose a new normal for a different type of topic, which nevertheless has potential bearings on all of the above topics, namely how we see artificial intelligence (AI) and the ethical issues that are linked to it. In the next section I will propose the use of the metaphor of an ecosystem to describe AI and explain why this is a suitable perspective. The final section will then look at the challenges that adopting this perspective entails.

2   AI as an Ecosystem

The idea to describe AI as an ecosystem is not new. References to AI ecosystems can frequently be found, in particular in policy-oriented documents, such as [1–3]. The attractions of this approach are easy to see. AI is not a monolithic technology, but rather a family of socio-technical systems that share some characteristics, notably the use of some digital technologies, most prominently machine learning. The ecosystems metaphor can accommodate the fact that there are many different and changing members of the systems and that there are nested systems of different sizes (e.g. organizational, national, international) that share this socio-technical nature. Ecosystems are furthermore not easily governed, as the relationships between their members tend to be complex, non-linear and often unpredictable.

This unpredictable nature of AI ecosystems is a reason for using the metaphor when talking about the ethics of AI. AI ecosystems consist of many different human, technical and organizational stakeholders. They raise a multitude of possible ethical and social concerns, many of which have no simple and obvious solutions. Elsewhere we have tried to explore what the use of the ecosystems metaphor means for the ethics of AI [4–6]. From this work arise a number of proposals of how AI ecosystems could be shaped in a way that would be conducive to human flourishing. We developed recommendations around three main areas: the delimitation of ecosystems, governance structures of ecosystems and the knowledge base of ecosystems. We propose that these different recommendations, taken together, offer the opportunity to better identify and address ethical and social concerns about AI.

3   Ethics and Systems – Future Challenges

While the use of the metaphor of an ecosystem provides a useful way of thinking about AI and its ethical issues, it also raises some new concerns. Ethical theory traditionally does not deal well with systems. Most approaches to ethics focus on the individual human being as the place of ethics and moral responsibility. Some suggestions have been made that clearly structured organizations might count as moral agents as well [7]. But loose agglomerations of agents such as those we find in an ecosystem are typically not covered by ethical theory.

This is a problem for moral philosophy, but it is one that translates into a problem of the practice of dealing with the ethics of AI. If ethics requires an individual rational agent but the ethical consequences of AI arise due to systems-based interaction, then there is no simple way of addressing these ethical issues.

It is therefore worth thinking about what an ethics of large socio-technical systems would look like. This calls for conceptual developments but also for empirical research to better understand how ethics is perceived and enacted in practice. This research programme will not only be important for dealing with the ethics of AI, but will have broader repercussions for the ethics of emerging technology more broadly. In light of the ever-growing importance of such emerging technologies for all aspects of human life, I believe that this research programme should be adopted with high priority.

References

1. European Commission: White paper on artificial intelligence: a European approach to excellence and trust. Brussels (2020)
2. UNESCO: First draft of the recommendation on the ethics of artificial intelligence. UNESCO, Paris (2020)
3. UK Government: national AI strategy (2021)
4. Stahl BC: Artificial intelligence for a better future: an ecosystem perspective on the ethics of AI and emerging digital technologies. Springer international publishing (2021)
5. Stahl, B.C.: Responsible innovation ecosystems: Ethical implications of the application of the ecosystem concept to artificial intelligence. Int. J. Inf. Manag. **62**, 102441 (2022). https://doi.org/10.1016/j.ijinfomgt.2021.102441
6. Stahl, B.C., Andreou, A., Brey, P., et al.: Artificial intelligence for human flourishing – beyond principles for machine learning. J. Bus. Res. **124**, 374–388 (2021). https://doi.org/10.1016/j.jbusres.2020.11.030
7. French, P.: Individual and Collective Responsibility. Schenkman, Cambridge, Mass (1972)

Managing Health Data Using Artificial Intelligence

Mirjana Ivanovic (iD)

Faculty of Sciences, University of Novi Sad, Novi Sad, Serbia
mira@dmi.uns.ac.rs

Abstract. Nowadays significant number of people suffer from a range of diseases. Also, population is getting older and older. It causes emergent need to produce a variety of medical and healthcare services that will help them to cope well with everyday activities. In this short paper we will put light on important aspects of health data management in order to achieve better patients' treatments.

Keywords: Machine learning · Health data management · Complex health data

1 Introduction

Modern population is facing everyday stressful situations, tense communication with other people, but also numerous unhealthy habits. There are a lot of possible circumstances, global serious health diseases like Covid-19 pandemic, that badly influence individual's health conditions. These key factors significantly increase appearance of critical health problems like cardiovascular, cancer, neurological, and others. Accordingly different key medical players and stakeholders initiated development of a range of medical/health platforms, frameworks and services with aim to help sick people to keep and even increase their quality of life (QoL) and wellbeing [1].

To achieve better diagnosis, treatment and propose better medical interventions it is necessary to have adequate datasets that contain comprehensive info of patients' health status, about disease, follow-ups, nutrition data, activity data and so on. Health data management is first step for preparation useful datasets for training personalized predictive machine learning (ML) models that will be used in everyday practice and support better medical decisions. After achieving reliable predictive models, it is necessary to present obtained results to physicians in more understandable and friendly way using techniques of data visualization and explainable artificial intelligence [1].

In the rest of the paper, we will briefly present key steps in processing big health data and some aspects of influences of AI techniques to support medical decisions.

2   Health Data Management

Monitoring health conditions is necessary to help physicians to better maintain patient's wellbeing. It is also essential step towards better prevention in decreasing appearance of serious diseases. Nowadays almost unavoidable is trend of using different smart and wearable devices for collection additional data of patient's activities [2] like weight, heart rate, number of steps, calories burned, sleep stages, and so on. Such additional sources of patient's data can increase reliability of developed predictive models.

To process patients' health records stored in databases and enhanced with additional sources like data from wearable devices, nutrition data, environmental data different health terminologies and ontologies have been used to achieve great level of standardization in data representations. These approaches support higher level of interoperability between different health information systems, services and data sources. Apart from necessary processes of standardization it is required to prepare datasets in forms suitable for processing using advanced techniques of artificial intelligence (AI) and ML.

The process of data aggregation is important step in producing compact patients' health records combining information from multiple sources. Individual patient's data are useful for physicians to make more effective treatment decisions, but they are limited only on the particular case. However, possibility to aggregate health data of numerous similar cases offers to a physician better understanding of status of a patient and helps to make better decisions. Aggregate management is crucial in patients' treatment as offers important insight in his/her health status [3].

Privacy preserving [4] is specific requirement in all situations when sensitive patients' data are exposed and should be used within health platforms, systems, and services. To prevent discovering sensitive patients' data at any stage of processing and using achieved results there are numerous techniques for de-identification like: Character/record masking, Shuffling, Anonymization, Collectively de-identification, Pseudonymization, Generalization. Apart from some widely use privacy preserving techniques like K-anonymity, L-diversity, and T-Closeness nowadays techniques as Differential privacy and Homomorphic Encryption plays essential role in big Cloud/Edge health architectures for processing complex health data [5].

Differential privacy is the systematic randomized modification that can be applied on dataset or on algorithm used for processing data in order to reduce information about the single patient. Homomorphic Encryption is security preserving technique that enables arithmetic operations on ciphertexts without need to decrypt original data. It is popular way to protecting sensitive patients' data leakage in distributed environments.

3   Health Data – AI/ML Models Development – Use of Models

From the point of view of application of powerful AI and specifically ML methods for processing big health datasets it is very important to have good quality patients' data. Incorrect and unclean data usually lead to wrong results and consequently negatively influence health decisions. For successful processing and use of health data several consecutive steps should be followed: Understanding of particular health domain and problem to be tackled; Think about ML aspects and collect necessary data; Data cleaning in order to correct inconsistencies and deal with missing values; Feature engineering that will help in selecting most influential features within available data-types; Train predictive models, evaluate them and use them for predictions; Visualize predictive models results in friendly form to be presented to physicians and health stakeholders.

Essential step when health data is cleaned and properly aggregated is building predictive models. Nowadays very popular approaches are Neural networks and Deep learning. These techniques reach very good performances however their results are usually hard to be interpreted/visualized. Therefore, simpler classification and regression techniques which are also very powerful in processing health date are applied. Their advantage is that results can be easily visualized and presented using explainable AI.

Usual classification learning algorithms used for processing health data are: Naive Bayes, K-nearest neighbors, Decision Trees, (bagging) Classification and Regression Trees, C5.0, Random Forests, Logistic Regression, Artificial Neural Network, Support Vector Machines and Linear Discriminant analysis. Characteristics regression approaches that are usually used for predicting numeric values, specific interventions or continuous variable are: Linear regression, Ridge regression, Lasso regression, Elastic net regression, Kernel ridge regression, regression by Support vector machines, Regression by random forests, and K-nearest neighbors [1, 6].

4   Conclusion

Rapid technological development, stressful style of life cause higher appearance of serious and chronic diseases. On the other hand, individual's QoL and wellbeing are getting to be more and more important. Emergent technologies and powerful AI techniques supports efficient big health data processing and offers better health services and more reliable medical decisions that help in increasing patient' QoL [7]. Unfortunately, personalized medicine has some limitations [6] that should be solved in order to support reliable use in everyday practice in hospitals and healthcare institutions.

References

1. He, J., Baxter, S.L., Xu, J., Xu, J., Zhou X., Zhang K.: The practical implementation of artificial intelligence technologies in medicine. Nat. Med. **25**, 30–36 (2019)
2. Burmester, G.R.: Rheumatology 4.0: big data, wearables and diagnosis by computer. Ann. Rheum. Dis. **77**(7), 963–965 (2018)
3. Lahiri, C., Pawar, S., Mishra, R.: Precision medicine and future of cancer treatment. Precis. Cancer Med. **2**, 33 (2019)
4. Siddique, M., Mirza, M.A., Ahmad, M., Chaudhry, J., Islam, R.: A survey of big data security solutions in healthcare. In: International Conference on Security and Privacy in Communication Systems, pp. 391–406. Springer, Cham. (2018). https://doi.org/10.1007/978-3-030-01704-0_21
5. Kaissis, G.A., Makowski, M.R., Rückert, D., et al: Secure, privacy-preserving and federated machine learning in medical imaging. Nat. Mach. Intell. **2**, 305–311 (2020). 10.1038/s42256-020-0186-1
6. Ivanovic, M.: Role of artificial intelligence in medical predictions, interventions and quality of life. In: 7th International Conference on Systems and Informatics (ICSAI), pp. 1–4 (2021). 10.1109/ICSAI53574.2021.9664199
7. Ray, P.P., Dash, D., De, D.: A systematic review of wearable systems for cancer detection: current state and challenges. J. Med. Syst. **41**(11), 180 (2017)

Contents

Keynote Talk

The New Normal: Innovative Informal Digital Learning After
the Pandemic 3
   John Traxler

Theoretical Foundations and Distributed Computing

StegIm: Image in Image Steganography 13
   Ivo Tasevski, Jovana Dobreva, Stefan Andonov, Hristina Mihajloska,
   Aleksandra Popovska-Mitrovikj, and Vesna Dimitrova

A Property of a Quasigroup Based Code for Error Detection 26
   Natasha Ilievska

Multi-access Edge Computing Smart Relocation Approach from an NFV
Perspective 38
   Cristina Bernad, Vojdan Kjorveziroski, Pedro Juan Roig,
   Salvador Alcaraz, Katja Gilly, and Sonja Filiposka

Artificial Intelligence and Deep Learning

MACEDONIZER - The Macedonian Transformer Language Model 51
   Jovana Dobreva, Tashko Pavlov, Kostadin Mishev, Monika Simjanoska,
   Stojancho Tudzarski, Dimitar Trajanov, and Ljupcho Kocarev

Deep Learning-Based Sentiment Classification of Social Network Texts
in Amharic Language 63
   Senait Gebremichael Tesfagergish, Robertas Damaševičius,
   and Jurgita Kapočiūtė-Dzikienė

Using Centrality Measures to Extract Knowledge from Cryptocurrencies'
Interdependencies Networks 76
   Hristijan Peshov, Ana Todorovska, Jovana Marojevikj,
   Eva Spirovska, Ivan Rusevski, Gorast Angelovski, Irena Vodenska,
   Ljubomir Chitkushev, and Dimitar Trajanov

Applied Artificial Intelligence

Evaluating Micro Frontend Approaches for Code Reusability   93
  Emilija Stefanovska and Vladimir Trajkovik

Combining Static and Dynamic Features to Improve Longitudinal Image
Retrieval for Alzheimer's Disease   107
  Katarina Trojachanec Dineva, Ivan Kitanovski, Ivica Dimitrovski,
  Suzana Loshkovska, and for the Alzheimer's
  Disease Neuroimaging Initiative

Architecture for Collecting and Analysing Data from Sensor Devices   121
  Dona Jankova, Ivona Andova, Merxhan Bajrami, Martin Vrangalovski,
  Bojan Ilijoski, Petre Lameski, Katarina Trojachanec Dineva,

Education

Adapting a Web 2.0-Based Course to a Fully Online Course
and Readapting It Back for Face-to-Face Use   135
  Katerina Zdravkova

Challenges and Opportunities for Women Studying STEM   147
  Mexhid Ferati, Venera Demukaj, Arianit Kurti, and Christina Mörtberg

Medical Informatics

Novel Methodology for Improving the Generalization Capability
of Chemo-Informatics Deep Learning Models   161
  Ljubinka Sandjakoska, Ana Madevska Bogdanova, and Ljupcho Pejov

An Exploration of Autism Spectrum Disorder Classification
from Structural and Functional MRI Images   175
  Jovan Krajevski, Ilinka Ivanoska, Kire Trivodaliev,
  Slobodan Kalajdziski, and Sonja Gievska

Detection of High Noise Levels in Electrocardiograms   190
  Danche Papuchieva and Marjan Gusev

Author Index   205

Keynote Talk

The New Normal: Innovative Informal Digital Learning After the Pandemic

John Traxler[✉] ⓘD

UNESCO Chair, Llanon, UK
John.Traxler@wlv.ac.uk

Abstract. The global pandemic catalysed a large-scale shift across many sectors of education towards digital learning. This was motivated by the need to preserve the ongoing delivery of education. Consequently, it was conservative rather than innovative, reinforcing existing face-to-face pedagogies rather than challenging them, and only benefitting learners actually within the sectors of education, further increasing the disadvantage of those not in the education sectors. This chapter outlines some innovative informal digital learning pedagogies that could be freely exploited to contribute to a more equitable 'new normal'.

Keywords: Informal digital learning · New normal · Innovative pedagogies

1 Introduction

This contribution builds on several observations and then outlines pedagogies that educators should consider as the world gradually comes out of the covid-19 pandemic. The pedagogic ideas are intended to exploit free web2.0 and social media applications and learners' existing confidence and familiarity. The observations embrace:

- The pandemic produced a global 'pivot' to digital learning in every sector of formal education, however

    - this 'pivot' was largely conservative, merely extending existing platforms and materials, merely transferring face-to-face pedagogies and merely deploying established faculty, owing to operational urgency, risk avoidance and resource constraints
    - this 'pivot' took only took place within formal education thereby disadvantaging people and communities excluded, ignored or oppressed by education systems, increasing their disadvantage

- People, communities and cultures often have their own traditions of learning different or distant from national or international norms, resources and organisations.
- People, communities and cultures all have their own experiences of creating, sharing, discussing, valorising, transforming, discarding and consuming the images, ideas and information they value and have meaning.

K. Zdravkova and L. Basnarkov (Eds.): ICT Innovations 2022, CCIS 1740, pp. 3–10, 2022.
https://doi.org/10.1007/978-3-031-22792-9_1

- Mobiles, mostly, and social media, meaning web2.0, mostly, are accessible, familiar, 'owned', controlled by everyone.
- Innovative digital learning has a range of emerging pedagogies that could be appropriated, adapted and integrated including, heutagogy, learner-generated content, badges, game mechanics, e-portfolios, curating resources, blogging, flipped learning ….

These observations provide the context for developing the following collection of innovative informal digital pedagogies. They can be deployed across a range of free social media applications; whichever social media applications are most widely used in any given community. They might include Facebook, Twitter, YouTube, LinkedIn, WordPress, Instagram, Wikipedia and Diigo. The policy focus is capacity-building and staff development not procurement and management, building autonomous, creative and critical community learning instead of passive institutional consumption.

1.1 Curation of External Resources

The digital world is awash with digital resources, many of them 'free' or 'open' and easily available. These resources include:

- content, for example websites, podcasts, downloads, audio, e-books, video, documents, images and text,
- communities, for example listservs, mail-bases, special interest groups, and, to give a concrete examples, Groups in Facebook and LinkedIn,
- tools, for example, for creating content, building communities or connecting people, using the keyboard, the microphone and the camera.

Educators and learners must develop the skills to exploit these resources [1]. These skills include not only the technical skills of online searching, classification and storage but also the critical skills of judging the provenance, purpose, credibility, stability, value and assumptions of the resources. They are the skills of 'curation' or 'orchestration'.

Curation encourages identifying and articulating the key arguments, issues or assets of a resource – is it any good, is it any better, why is it useful, is it original, is it permanent? Curation encourages classification and organization – how should a resource be described, what is its metadata, attributes or characteristics, what folksonomy, meaning what learner-generated organisation, classification or structure, works best for our cadre, our community or our school?

There are nevertheless reasons for caution, including the predominance of resources that embody American English culture and language, raising concerns about fragile, marginal, endangered or indigenous languages and cultures and concerns about the exact nature of both 'open' and 'free'.

Many 'open' educational resources in practice have highly formal and institutional characteristics, and their repositories and metadata reflect Western European ways of thinking about knowledge, education and learning. This makes searching and finding challenging. On the other hand, resources, having been found, can be organised, that is 'curated' in ways that reflect the learners' own knowledge structures using 'tagging' and 'folksonomies' [2, 3], using easy and lightweight tools such as Diigo [4] to create and organise individual or community libraries, a process sometimes known as 'social book-marking'.

Seen from a different perspective, curation could be viewed as a flipped classroom activity [5, 6] where learners are encouraged to find and critique resources before regrouping in the digital classroom to discuss and compare them.

1.2 Learner Generated Resources

Learner-generate resources are the content, communities and tools that learners themselves produce. There are arguments complementary to those for curation, namely that learners learn best from resources to which they can most closely relate, resources produced by other learners like themselves, and that this process validates them as worthwhile participants and contributors in the learning of other learners [7]. They are no longer just consumers. These resources may be content, meaning learners can produce and contribute ideas and images; perhaps, stories, explanations, examples, opinions and anecdotes.

Learner-generated resources align particularly with cultures, traditions and communities that specifically learn from narratives, tales and anecdotes and align to practices of (digital) story-telling, (DST) [8, 9]. They also resonate with the nature of learning in citizen science [10, 11] and with the transmission and preservation of local, informal and indigenous knowledge [12, 13] and folk skills. For any group of learners, these could be shared assets, archived and tagged alongside curated external resources, endorsing their validity and value.

1.3 Game Mechanics

Information, ideas and opinions now emerge rapidly and change rapidly, and there is often no stable or authoritative source for much essential learning or training. So, it is important for communities of learners to be able to evaluate resources amongst themselves, to develop their own judgments. This involves developing the skills by which learners can critique and review digital resources and crucially can then critique and review the reviews of others in their community and understand each reviewers' strengths and weaknesses, the likes and dislikes, of colleagues within their community enabling the growth of a self-critical community, able to calibrate each other's judgements.

This is where we need to exploit the mechanics of games, those functionalities, such as points, badges [14], levels, leader boards, etc., that allow learners to compete, compare and collaborate. There is ongoing development, and some reviews [15–19] provide systematic frameworks linking learning and game attributes, outcomes, feedback and roles, including the kinds of levels and roles we describe here. This can be combined with the stars, likes and reviews more commonly associated with online retailing and other informal activities, for example Goodreads, iTunes, Duo Lingo, YouTube and Amazon, to give systems that allow learners to be more resilient, collaborative, creative, active, critical and autonomous.

1.4 Blogging

Blogging is not explicitly a pedagogy but might be considered as an outlet for learner generated writing, an online form of expression giving wider visibility to learners' ideas, experiences, opinions and perspectives, and thus potentially increased self-esteem and self-confidence [20, 21].

1.5 Project-Based Learning

Learners can be engaged and mentored in individual learning activities, generic and community ones. These might include

- citizen science,
- local history,
- natural history,
- personal reflection,
- family history,
- urban geography,
- creative writing,
- physical exercise,

or more serious ones such as developing community teaching resources, devising tests or quizzes, making community podcasts, videos or recordings for other learners; they would use cell phones, tethered tablets or networked laptops to log ideas and share ideas, information, resources, and data and then exploit an informal flipped learning approach, gathering online in order to discuss, compare and critique their findings, results and thoughts [22, 23].

1.6 Personal Learning Environments

Learners should appropriate, adopt and adapt those resources that match them and their needs and their preferences, regardless of their institutional or organisational resources, priorities, policies or provisions. This imperative is conceptualised as personal learning environments, PLE, as an antidote to the institutional momentum, focus and values of the default and widespread virtual learning environment, VLE [24, 25]. Again, this is a concept intended to enhance learner agency, self-efficacy and personalised lifelong learning.

1.7 Heutagogy

Heutagogy is also called self-directed learning, [26–28]. It comprises whatever learners need to decide, manage and control their own learning. It is based on ideals of self-efficacy, human agency, reflection, and metacognition. Learners are actively involved in their learning processes. They decide what they will learn. They decide how, where and when it will be learned. It takes a non-linear path, that is determined by the learners. Assessment is collaborative, decided upon between teacher and learner or perhaps amongst a peer-group, for example, through the use of so-called learning contracts, learner-directed questions, flexible curriculum, and learning based on project work [29]. Advocates of heutagogy talk about its 'double-loop' learning [30], a kind of meta-cognition, learners learning about their learning.

1.8 Mobile Learning

Mobile learning often currently means high-tech small-scale subsidised pilot projects [31,32]. It should mean whatever learning is adapted, adopted and appropriate to communities and countries characterised, each in their different individual ways, by substantial movement and connection in all manner of different forms [33]. This version of mobile lerning would enable and enhance the other pedagogies we describe here. Our emphasis is ownership and control of learning within communities rather those that are either imposed or supplied externally, from outside and from above.

1.9 Open Learning

Open learning is the ideal that there should be no barriers to learning, and that organisations, for example teachers, authors, publishers, colleges, schools, universities and ministries. It is a movement, and it is systems and practices. It should make learning available freely and easily, that they should make resources freely available without restrictions on copying, adaptation, publishing and distribution.

Open Educational Resources are one specific realization of open learning (OER) [34]. [35], often housed in freely-accessible repositories. They are the most mature aspect of open learning, but the idea embraces teaching practices and delivery too.

1.10 E-Moderation

E-moderation is the set of techniques designed to incrementally transform a group of online learners, dependent on their tutor, into a self-reliant and self-sustaining group of learners functioning independently. These techniques have been set out for networked learning [36] as an easy five-step programme and adapted to a limited extent for mobile learners [37]. There are cultural dimensions to such a process, for example in moving the locus of control from the teacher and the institution to the learners and their group.

1.11 E-Portfolios

E-Portfolios are individual digital collections created by learners [38], of diverse and comprehensive aspect of their work, including posters, blogs, essays, websites, photographs, videos, podcasts and artwork. They may however also usefully capture other aspects of their life, such as volunteer experiences, family responsibilities, sporting achievements, employment history, community involvement, extracurricular activities. They record learning. They make it visible. Any worthwhile e-portfolio is two things, a product (meaning, a digital collection of artifacts) and a process (meaning, reflecting on them, the artifacts and what exactly the artifacts represent as learning and learning about learning).

e-Portfolios are able to provide the useful or necessary evidence for entry and progress in relation to employment, training or formal education [39, 40]. Within a community of practice, they enable critical and communal review of shared learner generated content. They can include and collate posts and contributions from across a portfolio of courses and they can create a sustainable community of practice, and also record episodes of micro-learning [41].

2 Conclusions

This contribution draws on research undertaken by the author and his team [42] for the Edtech Hub, established by the UK government. There was widespread concern that the measures used by governments worldwide to maintain the continued and ongoing delivery of education systems both during and after the pandemic would increase the disadvantage of those people, communities and cultures not within their country's education systems. There was also concern that these measures were driven by conservative managerial and operational pressures and that an opportunity for pedagogic innovation and transformation had been missed. This contribution outlines a few digital pedagogies that could be developed to support active lifelong learning alongside and outside formal education systems to significantly enhance critical, engaged and active community learning.

References

1. Mihailidis, P., Cohen, J.N.: Exploring Curation as a core competency in digital and media literacy education. J. Interact. Med. Educ. **2013**(1), 2 (2013). Part 2
2. Trant, J.: Studying social tagging and folksonomy: a review and framework. J. Digit. Inf. **10**(1) (2009). https://repository.arizona.edu/bitstream/handle/10150/105375/trant-studyingFolksonomy.pdf?sequence=1
3. Gupta, M., Li, R., Yin, Z., Han, J.: Survey on social tagging techniques. ACM SIGKDD Explor. Newsl. **12**(1), 58–72 (2010)
4. Diigo Homepage. https://www.diigo.com/. Accessed 20 Oct 2022
5. Bishop, J.L., Verleger, M.A.: The flipped classroom: a survey of the research. In: ASEE National Conference Proceedings, Atlanta, GA, vol. 30, no. 9, pp. 1–18 (2013)
6. Yough, M., Merzdorf, H.E., Fedesco, H.N., Cho, H.J.: Flipping the classroom in teacher education: implications for motivation and learning. J. Teach. Educ. **70**(5), 410–422 (2017)
7. Dyson, L.E.: Student-generated mobile learning: a shift in the educational paradigm for the 21st century. anzMLearn Trans. Mob. Learn. **1**(1), 5–19 (2012)
8. Robin, B.: The educational uses of digital storytelling. In: Society for Information Technology & Teacher Education International Conference, pp. 709–716. Association for the Advancement of Computing in Education (AACE) (2006)
9. Prins, E.: Digital storytelling in adult education and family literacy: a case study from rural Ireland. Learn. Med. Technol. **42**(3), 308–323 (2017)
10. Bonney, R., et al.: Citizen science: a developing tool for expanding science knowledge and scientific literacy. Bioscience **59**(11), 977–984 (2009)

11. Newman, G., Wiggins, A., Crall, A., Graham, E., Newman, S., Crowston, K.: The future of citizen science: emerging technologies and shifting paradigms. Front. Ecol. Environ. **10**(6), 298–304 (2012)
12. Kapuire, G.K., et al.: Technologies to promote the inclusion of Indigenous knowledge holders in digital cultural heritage preservation. In: International Conference on Culture & Computer Science (2016)
13. Maasz, D., Winschiers-Theophilus, H., Stanley, C., Rodil, K., Mbinge, U.: A digital indigenous knowledge preservation framework: the 7C model—Repositioning IK holders in the digitization of IK. In: Jat, D.S., Sieck, J., Muyingi, H.-N., Winschiers-Theophilus, H., Peters, A., Nggada, S. (eds.) Digitisation of Culture: Namibian and International Perspectives, pp. 29–47. Springer, Singapore (2018). https://doi.org/10.1007/978-981-10-7697-8_3
14. Ostashewski, N., Reid, D.: A history and frameworks of digital badges in education. In: Reiners, T., Wood, L.C. (eds.) Gamification in Education and Business, pp. 187–200. Springer, Cham (2015). https://doi.org/10.1007/978-3-319-10208-5_10
15. Lameras, P., Arnab, S., Dunwell, I., Stewart, C., Clarke, S., Petridis, P.: Essential features of serious games design in higher education: linking learning attributes to game mechanics. Br. J. Educ. Technol. **48**(4), 972–994 (2017)
16. Chorney, A.I.: Taking the game out of gamification. Dalhousie J. Interdisc. Manag. **8**(1) (2012). https://dalspace.library.dal.ca/handle/10222/16030?show=full
17. Kusuma, G.P., Wigati, E.K., Utomo, Y., Suryapranata, L.K.P.: Analysis of gamification models in education using MDA framework. Proc. Comput. Sci. **135**, 385–392 (2018)
18. Kim, B.: Game mechanics, dynamics, and aesthetics. Libr. Technol. Rep. **51**(2), 17–19 (2015)
19. Callaghan, M.J., McShane, N., Eguiluz, A.G., Teilles, T., Raspail, P.: Practical application of the Learning Mechanics-Game Mechanics (LM-GM) framework for Serious Games analysis in engineering education. In: 13th International Conference on Remote Engineering and Virtual Instrumentation, pp. 391–395. IEEE (2016)
20. Salen, T.: Weblogs and blogging constructivist pedagogy and active learning in higher education. Master's thesis. The University of Bergen (2007)
21. O'Donnell, M.: Blogging as pedagogic practice: artefact and ecology. Asia Pac. Med. Educ. **17**, 5–19 (2006)
22. Kokotsaki, D., Menzies, V., Wiggins, A.: Project-based learning: a review of the literature. Improv. Sch. **19**(3), 267–277 (2016)
23. Smith, M., Gurton, P.: Flipping the classroom in teacher education. In: Walker, Z., Tan, D., Koh, N.K. (eds.) Flipped Classrooms with Diverse Learners. STE, pp. 221–238. Springer, Singapore (2020). https://doi.org/10.1007/978-981-15-4171-1_13
24. Dabbagh, N., Kitsantas, A.: Personal Learning Environments, social media, and self-regulated learning: a natural formula for connecting formal and informal learning. Internet High. Educ. **15**(1), 3–8 (2012)
25. Wilson, S., Liber, O., Johnson, M., Beauvoir, P., Sharples, P., Milligan, C.: Personal Learning Environments: challenging the dominant design of educational systems. J. E-learn. Knowl. Soc. **3**(2), 27–38 (2007)
26. Blaschke, L.M.: Heutagogy and lifelong learning: a review of heutagogical practice and self-determined learning. Int. Rev. Res. Open Distrib. Learn. **13**(1), 56–71 (2012)
27. McLoughlin, C., Lee, M.J.: Personalised and self-regulated learning in the Web 2.0 era: international exemplars of innovative pedagogy using social software. Australas. J. Educ. Technol. **26**(1), 28–43 (2010)
28. Moore, R.L.: Developing lifelong learning with heutagogy: contexts, critiques, and challenges. Distance Educ. **41**(3), 381–401 (2020)
29. Blaschke, L.M., Hase, S.: Heutagogy, technology, and lifelong learning for professional and part-time learners. In: Dailey-Hebert, A., Dennis, K.S. (eds.) Transformative Perspectives and

Processes in Higher Education. ABET, vol. 6, pp. 75–94. Springer, Cham (2015). https://doi.org/10.1007/978-3-319-09247-8_5

30. Eberle, J.H.: Lifelong learning. In: Self-determined Learning: Heutagogy in Action, pp. 145–157 (2013)
31. Traxler, J.: Learning in a mobile age. Int. J. Mob. Blended Learn. **1**(1), 1–12 (2008)
32. Kukulska-Hulme, A., Traxler, J.: Mobile Learning: A Handbook for Educators and Trainers (Lockwood, F. (Series ed.)). Routledge, London (2005)
33. Traxler, J.: Learning with mobiles in the digital age. Pedagogika Spec. Monothematic Issue: Educ. Futures Digit. Age: Theory Pract. **68**(3), 293–310 (2018)
34. Butcher, N.: A Basic Guide to Open Educational Resources (OER). Commonwealth of Learning, Vancouver (2015)
35. Atkins, D.E., Brown, J.S., Hammond, A.L.: A review of the open educational resources (OER) movement: achievements, challenges, and new opportunities, vol. 164. Creative Common, Mountain View (2007)
36. Salmon, G.: E-Moderating: The Key to Teaching and Learning Online. Psychology Press, London (2003)
37. Brett, P.: Students' experiences and engagement with SMS for learning in higher education. Innov. Educ. Teach. Int. **48**(2), 137–147 (2011)
38. Xerri, D., Campbell, C.: E-portfolios in teacher development: the better option? ELT J. **70**(4), 392–400 (2016)
39. Wuetherick, B., Dickinson, J.: Why ePortfolios? Student perceptions of ePortfolio use in continuing education learning environments. Int. J. ePortfolio **5**(1), 39–53 (2015)
40. Heinrich, E., Bhattacharya, M., Rayudu, R.: Preparation for lifelong learning using ePortfolios. Eur. J. Eng. Educ. **32**(6), 653–663 (2007)
41. Buchem, I., Hamelmann, H.: Microlearning: a strategy for ongoing professional development. eLearning Papers **21**(7), 1–15 (2010)
42. Traxler, J., Scott, H., Smith, M., Hayes, S.: Learning through the crisis helping decision-makers around the world use digital technology to combat the educational challenges produced by the current COVID-19 pandemic (No. 1). EdTech Hub (2020)

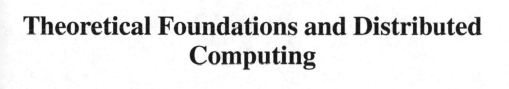

Theoretical Foundations and Distributed Computing

StegIm: Image in Image Steganography

Ivo Tasevski⍟, Jovana Dobreva⍟, Stefan Andonov⍟, Hristina Mihajloska⍟, Aleksandra Popovska-Mitrovikj(✉)⍟, and Vesna Dimitrova⍟

Faculty of Computer Science and Engineering, University Ss. Cyril and Methodius, Skopje, Republic of North Macedonia
ivo.tasevski@students.finki.ukim.mk,
aleksandra.popovska.mitrovikj@finki.ukim.mk

Abstract. Today, data security is a significant problem in data communication. Sometimes we do not realize how susceptible we are to cyberattacks and the theft of critical information, including everything, from social network passwords to complete identities. Any system, no matter how sophisticated it is, is vulnerable and susceptible to attack. Steganography is a technique for hiding secret information by adding it in a non-secret file. Also, it can be used in combination with encryption to further conceal or safeguard data. Therefore, in this paper, we present the StegIm model for hiding, retrieving, and detecting images using variants of LSB Encoding Steganography. We present the implementation of the model with different algorithms for hiding the image in another image (shape steganography algorithm and standard delimiter-based steganography algorithm) and give comparison analysis of used algorithms. To further strengthen the security, we implemented image encryption for this type of image steganography and analyzed the improvements, benefits, advantages, and disadvantages of this model in each phase of hiding/retrieving. Also, StegIm can detect hidden data in given images with the help of machine learning methods. This model will strive to reach the highest level of confidentiality, authenticity, and integrity of the user's confidential and sensitive data.

Keywords: StegIm · Crypt-steganography · Machine learning · Stego-image encryption

1 Introduction

Data security is regarded as a severe issue when data communication is carried out via the Internet, and everyone expects their data to be secure during the communication process. The process of concealing a message, image, audio, or video by embedding it in another message, image, audio, or video is known as steganography.

Steganography is a technique for hiding secret information by adding it in a regular, non-secret file or communication. The hidden information is subsequently extracted at the intended location. Steganography can be used to hide almost any type of digital content, including text, image, audio, or video content, and the data can be hidden in almost any other type of digital content. When we want to send some secret information,

© The Author(s), under exclusive license to Springer Nature Switzerland AG 2022
K. Zdravkova and L. Basnarkov (Eds.): ICT Innovations 2022, CCIS 1740, pp. 13–25, 2022.
https://doi.org/10.1007/978-3-031-22792-9_2

steganography can be seen as more discreet than cryptography, but in this case, the hidden message is much easier to find out.

Steganography can be combined with encryption to conceal further or safeguard data. The content we want to hide with steganography, known as hidden text, is often encrypted before being integrated into the carrier text file or data stream. This approach is known as crypt steganography. If the hidden text is not encrypted, it is commonly processed in some way to make it more difficult to detect the secret content.

There are numerous techniques to insert the hidden message into regular data files. One method is to conceal data in bits that correspond to consecutive rows of the same color pixels in a picture file. The output will be an image file that looks just like the original image but has "noise" patterns of regular, unencrypted data.

While there are many various applications for steganography, such as hiding sensitive data within certain file formats, one of the most widely used methods is to incorporate a text file within an image file. This is performed by putting the message with less significant bits in the data file so that anyone viewing the picture file should not be able to distinguish between the original image file and the stego-image. There are a lot of ways to hide information inside an image and some common approaches include Least Significant Bit Insertion, Masking and Filtering, Redundant Pattern Encoding, Encrypt and Scatter, Coding, and Cosine Transformation [1]. The pixels of an image consist of values R, G, and B for Red, Green, and Blue, respectively. Each value is in the 0–255 range and image's least-significant bits are the final bits of each byte The pixel's smallest value for each of the R, G, and B values are stored in the least-significant bits. In the Least Significant Bit (LSB) approach, these least significant bits of the image are changed [2]. Therefore, it is difficult to distinguish between the original and the stego-image.

On the other hand, steganalysis is the process of identifying the presence of hidden messages using steganography. This process uses methods for extracting hidden communication from stego-objects [3, 4]. For the detection of hidden data, in paper [5], authors present methods for steganography and steganalysis methodologies with machine learning frameworks. They show how machine learning frameworks can be used to uncover secret data concealed in images using steganography algorithms. In order to detect real-world stego-images, in [6] authors provide a deep learning system that employs Convolutional Neural Networks (CNN). Two pre-trained models with 73.33% and 79.43% accuracy are presented.

In paper [7] we propose a StegYou model for hiding, retrieving, and detecting digital data in images. The implementation of hiding and retrieving data is based on a custom-made crypt-steganography algorithm for one or more images. Also, StegYou can detect hidden data in given images using the Machine Learning approach and achieve an accuracy of 87% for Single-Image Steganography and accuracy of 98% for Single-Image Crypt-Steganography.

In this paper, we propose a modification of the StegYou model, called StegIm where we consider hiding and retrieving a secret image in a cover image. While steganography has many legitimate applications, malware developers have been found to use it to conceal the transmission of malicious code. Therefore, in our StegIm model, we incorporate a detection of hidden image in image using the machine learning methods.

The rest of the paper is organized in the following way. In Sect. 2 is given the implementation of StegIm for hiding and retrieving digital images using different encoding methods. In the same section is presented the implementation of the model with different algorithms for hiding the image in another image (image shape steganography algorithm and standard delimiter-based steganography algorithm). Also, we present comparison analysis of used algorithms and to further strengthen the security, we have also implemented image encryption for this type of image steganography. The improvements, benefits, advantages, and disadvantages of our model in each phase of hiding/retrieving are analyzed. The detection of hidden data in images for the StegIm model, using the machine learning method, is described in Sect. 3. Finally, in Sect. 4, we give some conclusions and further work on the proposed StegIm model.

2 Implementation of StegIm for Hiding and Retrieving Digital Images

In this section, we consider the implementation of StegIm for hiding and retrieving digital images. Now we will dive into the concept of hiding an image inside another image. Still, before we do, we should again mention some noteworthy things relevant to this concept of image steganography. As it is already known, a digital image can be represented as a finite set of digital values called pixels. Pixels are the smallest units of the digital image that can be displayed, and they contain values representing digital numeric RGB color data. Therefore, we can see an image as a two-dimensional array - a matrix of pixels that includes a fixed number of rows and columns. Each pixel has three decimal R, G, and B values, and each RGB value can be represented in binary format with 8 bits. In this section, we have defined three encoding methods, each with its own unique characteristics, benefits, and downsides. Since we are working with more extensive and denser data (images), it is crucial to have enough encoding capacity. This is the primary principle of the L4SB and L2SB encoding methods. As the name indicates, the L4SB encoding method uses the last four rightmost bits or last four LSBs (least-significant bits) to encode a secret image, while the latter uses the last two LSBs. This increases the encoding capacity and brings more data to be encoded in the carrier images. The third one is the standard LSB encoding method used in steganography, where data is hiding in images and use the last LSB of the carrier image for secret message encoding.

2.1 The LSB Encoding Method

The LSB encoding method is used to encode secret image into a carrier image, using the last rightmost bit of the carrier image's pixel. It allows us to encode a secret image which is eight times smaller than the carrier image. A carrier image of size N bytes can hold a secret image of at most N/8 bytes since we are encoding one R, G, or B value of the secret image inside eight R, G, and B values of the carrier image. This encoding method does not compromise the quality of the carrier stego-images. Also, the image quality loss is not visible to the bare eye, meaning that stego-images are very hard to detect, making this encoding method the safest and best in terms of quality retention. The encoding capacity is its biggest problem, but it can be surpassed by simply using bigger carrier images or smaller secret images.

2.2 The L2SB Encoding Method

The L2SB encoding method allows us to encode a secret image that is a fourth of the size of the carrier image in bytes, at most. A carrier image of N bytes can hold a secret image of nearly N/4 bytes, since we are encoding one R, G or B value of the secret image inside four R, G, and B values of the carrier image. This encoding method delivers a twice bigger encoding capacity than the previous one. It doesn't compromise the quality of the carrier stego-image, but some of the changes still can be visible to the bare eye. The L2SB encoding method breaks the ice between the L4SB and LSB encoding methods and strikes a compromise between the two when viewing from image quality and encoding capacity point of view.

2.3 The L4SB Encoding Method

Since we are encoding data in the last four bits of the values of the pixels in the carrier image, the L4SB encoding method allows us to encode an image that is half the size of the carrier image in bytes at most. Suppose an image has N values (every pixel contains three values, and each value is one byte). In that case, an image of nearly N/2 bytes can be encoded in such a carrier image since we are technically replacing half of the bits of the carrier image with the bits of the secret image. If nothing else needed to be encoded, then the encoding capacity would be exactly N/2 bytes.

The L4SB encoding method still follows the LSB encoding principles, with the only difference being that it is a modified version of it. Instead of using only the last LSB, it uses the last four LSBs, with the sole purpose of increasing the encoding capacity to fulfill the needs of encoding more dense and larger data formats, like JPG or PNG images.

However, it is crucial to mention that this encoding method drastically compromises the quality of the carrier stego-image, making it easily detectable and thus making the anonymity and privacy of secret images obsolete. It is the most insecure of all three encoding methods and simultaneously makes the whole steganography concept pointless.

The L4SB encoding method is very risky, and we do not recommend using it for textual or other smaller data formats.

To summarize, in Table 1, we give all advantages and disadvantages of previously presented encoding methods.

2.4 Used Algorithms for Hiding/Retrieving Images

In this subsection, we compare two algorithms for hiding an image inside another image - the image shape steganography algorithm and the standard delimiter-based steganography algorithm, which both carry unique principles and features. But, before we begin the analysis, we need to clarify a few things and explain some key terms used in the following.

A carrier image is an image in which the secret image is encoded/hidden, i.e., the image that carries our secret JPG or PNG data. A carrier image is the same as a stego-image; to be more precise, the carrier image is the stego-image after the image has been

Table 1. Comparison analysis of LSB, L2SB and L4SB encoding methods

Encoding method	Advantages	Disadvantages
LSB	Safest encoding method - carrier stego-image quality is slightly compromised, but not visible to the bare eye	Smallest encoding capacity - twice smaller than the one when using L2SB, and four times smaller than the one when using L4SB
L2SB	Encoding capacity is still big - twice bigger than the one when using the LSB encoding method Compromise between quality retention and encoding capacity	Carrier stego-image quality is slightly compromised, but visible to the bare eye Encoding capacity is twice smaller than the one when using L4SB
L4SB	Biggest encoding capacity - two times bigger than the one when using L2SB, four times bigger than the one when using LSB	Most insecure encoding method - carrier stego-image quality is drastically compromised Stego-image is easily detectable

encoded. Image shape represents the dimensions of an image, i.e., the height and width of an image. A shape flag is an indicator for the encoded image shape, done in the first two, four, or eight pixels of the carrier image, depending on the encoding method used. The shape flag is used to carry the dimensions of the secret image, just in order to make decoding process easily.

Essential functions used in the algorithms are the functions encode (for hiding data) and decode (for retrieving data). These functions are used as entry points to the actual functions that represent hiding in and retrieving the image from an image. Their variation depends on which algorithm is used for producing the stego-image. In the following we will give a brief explanation of the used algorithms.

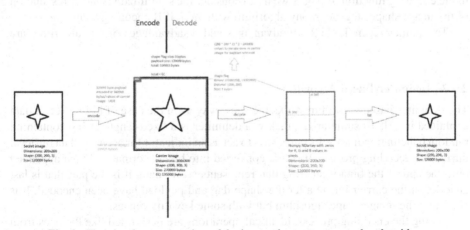

Fig. 1. The visual representation of the image shape steganography algorithm

2.4.1 Image Shape Algorithm

The core concept of this method lies purely in the image shape, i.e., the image dimensions of the secret image that are hidden in the carrier image. Knowing the dimensions of the secret image during the decoding process helps us find the exact shape of the secret image. It also opens up the possibility of extracting only the relevant information from the carrier image by calculating the exact number of values used for encoding. This is the core concept of our image shape algorithm.

This algorithm works by encoding the image shape in the carrier image's first two, four, or eight pixels, depending on the encoding technique. According to the fact that the encoding capacity of a pixel of a carrier image is 12 bits at most when using the L4SB encoding method, 6 when using the L2SB encoding method, and 3 when using the LSB encoding method, we decided to use 12 bits for both the height and width of an image to be encoded, resulting in maximum width/height of a secret image to be 4095. An image with dimensions 4095 × 4095 is the maximum allowed size of a secret image. Following the shape flag is the actual payload, i.e. the secret data that we encode in the carrier image. During the encoding process, the first step is determining the secret image dimensions. After that, the secret image is converted into binary format, i.e., the values of the pixels of the secret image for width and height are converted to binary and then concatenated together. The shape flag is encoded in the first two, four, or eight pixels of the carrier image, after which the binary bitstring of concatenated, flattened binary values are encoded in the carrier image. During the decoding process, the shape flag is extracted from the carrier image first, and the secret image size is calculated. By doing this, we now know the exact width and height of our secret image and the number of relevant R, G, and B values of the carrier image that have data encoded in them. Afterward, we iterate over the exact number of pixels and values of the carrier image and retrieve our payload, i.e., the bits of our secret image. The next step is grouping the extracted bits into groups of 8, representing bytes, and converting them into decimal format. Using the extracted width and height of the secret encoded image, we can reconstruct the exact secret image passed in the encoding function. In Fig. 1 we try to consider the simplified visual representation of the image shape steganography algorithm with the L4SB encoding technique.

To summarize, in Table 2, all advantages and disadvantages of this algorithm are given.

2.4.2 Delimiter-Based Algorithm

The delimiter-based algorithm works the same way as in the case of hiding secret data explained in [7]. To summarize quickly, a delimiter is a fixed-length string containing random punctuation marks. Its purpose is to act as an indicator for accurate data retrieval during the decoding process. It is first converted into binary format and then appended onto the end of the binary bitstring that represents secret data. It is the part that is last encoded in the carrier image after the shape flag and payload have been encoded. It is similar to the image shape algorithm but with some key differences.

During the encoding process, identical operations are performed like the ones from the previous algorithm, except that additional 10 bytes are added to the payload, corresponding to the delimiter. Firstly, the shape flag is encoded in the first two, four, or

Table 2. Advantages and disadvantages for the image shape steganography algorithm.

Advantages	Disadvantages
• Slightly bigger encoding capacity and simpler workflow • Only data from relevant R, G and B values from carrier image is extracted - relevant value calculation using shape flag • No irrelevant, unimportant data extracted - speed and efficiency • Faster when it comes down to small secret images and large carrier images	• Carrier stego-image quality might be drastically compromised • Secret image size must be up to eight times smaller than carrier image, depends on encoding technique used • Non-stego images can't be detected

eight pixels of the carrier image (depending on the encoding technique), followed by the payload (the secret bits of the secret image and the delimiter). Since the delimiter is located at the end of the payload, it is encoded last.

During the decoding process, the shape flag is first extracted from the carrier stego-image. After that, the process is identical to the one in the previous phase, except for the part where every extracted byte, apart from being converted to decimal, is converted to its corresponding ASCII characters. This is done for delimiter detection, and once the last ten extracted characters match the delimiter, the delimiter is removed from the final extracted values, leading to valid image retrieval. A simplified visual representation of the delimiter-based workflow with the L4SB encoding method is given in Fig. 2.

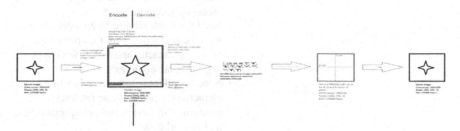

Fig. 2. The visual representation of the delimiter-based steganography algorithm.

To summarize, in Table 3, all advantages and disadvantages of this algorithm are given.

2.4.3   General In-Depth Analysis

According to Table 2 and Table 3 from the previous algorithms, we can conclude that using one of the algorithms has its own benefits.

Using an image shape algorithm makes the decoding process easy, but it lacks in the way that any image can be passed to the decoding function, and there is not a way to determine if that image is truly a stego-image or not. The decoding function will throw an error in most cases like these because of inconsistent values, but not for all. Also, an image shape algorithm is carrying the image size in the carrier image, so it is the best practice not to decode the whole stego-image and improves the implementation performances.

However, using and encoding a delimiter in the carrier image, stego-images can be easily detected when passed to the decoding function. In the previous case, the user could send any image and whether that image is a stego-image or not, there wasn't a way to determine that. Here, passed images can instantly be classified as stego or non-stego images simply by checking the presence of the delimiter in the image. This eliminates those very small number of cases where the user could pass a non-stego image and successfully decode data from it, even if it is not a stego-image. The delimiter-based algorithm will eliminate the presence of non-stego images during the decoding process, which makes it in a way a bit more reliable than the other algorithm. Nevertheless, that doesn't make it better than the image shape algorithm. This algorithm is inefficient to use, makes the workflow slightly more complex and might drastically increase processing time. With that said, the image shape algorithm takes the crown for this type of image steganography.

Table 3. Advantages and disadvantages for the delimiter-based steganography algorithm.

Advantages	Disadvantages
• Non-stego images can be detected • It does not depend on the size of the secret image	• Carrier stego-image quality might be drastically compromised • Secret image size must be up to eight times smaller than carrier image, depends on encoding technique used • Need for a delimiter to be encoded along with the secret data - uses carrier image encoding capacity and makes workflow slightly more complex • Need for all least-significant bits to be extracted for secret image retrieval - creation of irrelevant and unimportant data, makes workflow

Also, the most important metric for stego-images is the Peak Signal-to-Noise Ratio (PSNR) [8] as a way to prove the quality of a stego-image representation. The distortion in the stego-image can be measured by PSNR, and a higher value of PSNR indicates a lesser image distortion. We did an experiment with 400 color carrier images with different sizes and just one secret image of size 100×100 pixels. We choose to work with some popular image sizes like one for Viber image (1600×1200), Facebook and Instagram post images (1080×1080), Instagram Story (1080×1920) and Twitter post image (1200×675) [9]. We computed PSNR numbers to all stego-images obtained with the image shape algorithm, and the delimiter-based algorithm. All the stego-images are encoded using the previously explained encoding methods (LSB, L2SB, and L4SB). We found that some of the stego-images encoded with L4SB method had a PSNR number less than 50, so we decided to remove them from further analyses.

Table 4, shown the average values of PSNRs obtained by the three encoding methods on 400 colored images using the image shape algorithm, and the delimiter-based algorithm. The experimental results showed that LSB method has higher peak signal to noise ratio compared to others for both algorithms.

2.5 Image Encryption

To further strengthen the security, we have also implemented image encryption with authentication for this type of image steganography. We are using the PyNaCl python library for symmetric encryption with authentication. This library uses 32-byte symmetric keys, along with a 24-byte nonce and makes use of the XSalsa20 stream cipher [10], which has been proven to be a reliable and fast stream cipher. It also used Poly1305 MAC [11] authentication for data integrity and authenticity. Since we cannot encrypt the secret image directly, we need to first convert it to a byte-array. After this, we can encrypt the resulting byte-array using PyNaCl, by passing in a custom-generated nonce and symmetric encryption key. The output of the authenticated encryption process is another cipher byte-array, which contains the cipher-text, nonce and authentication data. We then convert that cipher byte-array into binary format by using our custom function and encode the resulting bitstring into our carrier image. During the decoding process, after the data is extracted, the ASCII decimal representation of every extracted byte is retrieved and stored in a byte-array. The byte-array is then decrypted using PyNaCl and the previously generated symmetric key and nonce. After the decryption is over, we load the resulting, decrypted byte-array and return the data and fully retrieve the secret image.

Table 4. PSNR analysis for image shape and delimiter-based algorithms

Encoding method	PSNR (average value)	
	Stego-image obtained with image shape alg.	Stego-image obtained with delimiter-based alg.
LSB	63.55	63.55
L2SB	59.57	59.61
L4SB	51.83	51.85

It is important to mention that image encryption does not work with the image shape algorithm, since the size of the resulting cipher byte-array is dynamic and diverse when it comes down to secret images of different sizes and dimensions. Thus, image encryption only works with the delimiter-based algorithm.

There is still an option to not use image encryption at all. The code has been adjusted and optimized to enable selection of desired algorithm and encryption, by simply passing additional parameters to the encoding and decoding functions. The reason for this is that we don't need the secret image dimensions for image retrieval and reconstruction when we use image encryption, since we retrieve byte data that already has that information encoded in it. When Pickle deserializes that data, the secret image is automatically reconstructed and returned, leaving us with the opportunity of not having to deal with an additional property, that being the shape flag.

3 Machine Learning Implementation of StegIm for Detecting Digital Data in Images

Our StegIm model for detection of hidden images in images is based on the Machine Learning approach, where with the help of supervised learning we train several machine learning models for recognizing stego and non-stego images that were created with the usage of various carriers. We are training and evaluating different classification models, with the purpose of discovering the image dimensions and stego encoding algorithms which produce the best security for hiding information.

We conducted 4 experiments using the LSB encoding method for the following resolutions of the carrier images:

- 1080×1080 pixels (resolution used in posts on Facebook and Instagram)
- 1080×1920 pixels (resolution used in stories on Instagram)
- 1200×675 pixels (resolution used in posts on Twitter)
- 1600×1200 pixels (max resolution used on Viber for sharing images).

In Subsect. 3.1 we describe the phase of data sets generation and preprocessing. In Subsect. 3.2 we present the machine learning model architecture that resulted in the best results while in Subsect. 3.3 we present the best results from the evaluation of the chosen model.

3.1 Data Sets Generation and Preprocessing

For all previously defined 4 experiments, we created different datasets. Each dataset corresponds to a different carrier resolution as defined above. We used 100 carriers in all 4 experiments and accordingly 100 secret images with a resolution of 100×100 pixels so that each secret image was hidden in a different carrier. After the encoding was done using the LSB encoding and the image shape algorithm, in a separate textual file we labeled all the stego images with a binary label 1. Lastly, we included in the dataset, the original 100 carriers that were used to hide images, with a binary label 0 (non-stego image).

After the datasets were created, we preprocessed each of the datasets with the help of the previously mentioned function for converting a given image into bits, so basically, we mapped all of the image datasets into a tabular dataset (which can be used in traditional ML approaches) where each column represents the last bit in each of the RGB channels of each pixel. Depending on the experiment here is given the list with the input binary vector sizes:

- Twitter Post Image: 2 430 000

- Facebook/Instagram Post Images: 3 499 200
- Viber Images: 5 760 000
- Instagram Story: 6 220 800.

In order to perform an evaluation of the trained models, we split each dataset into a training dataset (80% or 160 images) and a testing dataset (20% or 40 images). The split of the dataset is completely balanced i.e. the training dataset has 80 stego images and 80 non-stego images and the testing dataset accordingly has 20 stego and 20 non-stego images.

3.2 Model Architecture

To achieve exceptionally high prediction skills, the gradient boosting method [12] was developed. However, the algorithm's application has been constrained since it requires that just one decision tree be constructed at a time in order to minimize the errors of all earlier trees in the model. Because of this, training even the smallest models took a long time. Gradient boosting underwent a revolution when eXtreme Gradient Boosting (XGBoost) emerged as a cutting-edge method. In XGBoost, individual trees are constructed over many cores, and data is organized to speed up searches [13]. Model performance increased and model training time was cut in half consequently.

For our problem, we use the following hyper-parameters for XGBoost model:

- Maximal depth of the decision tree: 68
- Number of parallelized decision trees: 10
- Learning rate: 0.16.

3.3 Results

Due to hardware and software limitations as well as the huge sizes of the datasets defined for Viber images and Instagram stories, we could not complete the experiments and provide results in this paper.

The following metrics were used for evaluation: F1 score, Accuracy and Recall. The precision and recall of the test dataset are used to calculate the F1 score, with precision being the number of true positive results divided by the total number of positive results, and Recall being the number of true positive results divided by the total number of samples that should have been identified as positive. In Table 5, are shown the results from the evaluation of the StegIm model.

Table 5. Results from the Machine Learning approach

Image on social media posts	Accuracy	Recall	F1 score
Twitter	100%	100%	100%
Facebook/Instagram	97%	98%	97%

In Fig. 3 we present the confusion matrix, which represents the number of False positives, False negatives, True positives, and True negatives obtained with Twitter Image posts and Facebook or Instagram Image posts. Where, in our case the positive class is the one labeled with **1**, in which the secret image is hidden.

From this we can conclude that the performance of our model is quite good for the original images with smaller dimensions, while the evaluation metrics for larger dimensions tend to decline. Therefore, we hope to train the classification model on larger images in the future along with the various image-in-image hiding algorithms. And, with this we will demonstrate through the evaluation metrics that image dimensions are important in the stego encoding algorithms.

4   Conclusion and Future Work

StegIm is a model for hiding and retrieving digital images into other images. It uses the LSB encoding method as commonly used in image steganography. Instead, we also use L2SB and L4SB encoding methods to increase the encoding capacity but still not compromise the quality of obtained stego-images. The actual parameter for distortion in stego-images is PSNR. The PSNR values obtained from the three methods with the given two algorithms (image shape and delimiter based) are very similar. Besides hiding, StegIm is used for retrieving the secret image from the stego-image. Also, we use the power of machine learning and implement the successful detection of the hidden image in one standard image (without knowing that it is a stego-image).

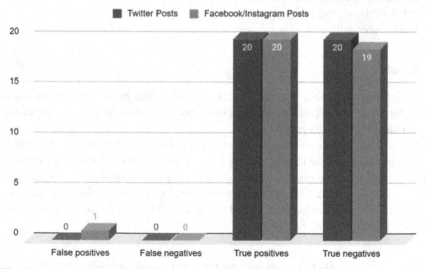

Fig. 3. Prediction of the StegIm model for Twitter, Facebook and Instagram Image Posts

Our plan for future work is to extend our work to other encoding algorithms and make our comparison analysis broader and more in-depth. Also, we will try to extend StegIm as a tool for hiding/retrieving secret image into multiple images, so that we can focus more on the security part of this methodology.

References

1. Bandyopadhyay, S.K., Bhattacharyya, D., Ganguly, D., Mukherjee, S., Das, P.: A tutorial review on steganography. In: International Conference on Contemporary Computing (IC3-2008) (2008). https://www.jiit.ac.in/jiit/IC3/IC3_2008/IC32008.html
2. Al-Azzeh, J., Alqadi, Z., Ayyoub, B., Sharadqh, A.: Improving the security of LSB image steganography. Int. J. Inform. Vis. **3**(4), 384–387 (2019). Computer Engineering Department, Al Balqa'a Applied University, Amman, Jordan
3. Babu, J., Rangu, S., Manogna, P.: A survey on different feature extraction and classification techniques used in image steganalysis. J. Inf. Secur. **8**, 186–202 (2017). https://doi.org/10.4236/jis.2017.83013
4. Oludele, A., Sunday, I., Afolashade, K., Uchenna, N.: Security test using StegoExpose on hybrid deep learning model for reversible image steganography. Int. J. Comput. Trends Technol. **70**(5), 7–14 (2022)
5. Ki-Hyun, J.: A study on machine learning for steganalysis. In: Proceedings of the 3rd International Conference on Machine Learning and Soft Computing (ICMLSC 2019), pp. 12–15. Association for Computing Machinery, New York, NY, USA (2019). https://doi.org/10.1145/3310986.3311000
6. Lavania, K.: A Deep Learning Framework to identify real-world stego images. Master's thesis. National College of Ireland, Dublin (2021)
7. Tasevski, I., Nikolovska, V., Petrova, A., Dobreva, J., Popovska-Mitrovikj, A., Dimitrova, V.: StegYou: model for hiding, retrieving and detecting digital data in images. In: Future Technologies Conference (FTC 2022), Vancouver, Canada (2022). Accepted
8. Wang, Z., Simoncelli, E.P., Bovik, A.C.: Multiscale structural similarity for image quality assessment. In: The Thirty-Seventh Asilomar Conference on Signals, Systems & Computers, vol. 2, pp. 1398–1402 (2003). https://doi.org/10.1109/ACSSC.2003.1292216
9. Social Media Image Sizes Guide. https://www.constantcontact.com/blog/social-media-image-sizes/. Accessed 01 June 2022
10. Bernstein, D.J.: Extending the Salsa20 nonce. http://skew2011.mat.dtu.dk/proceedings/Extending%20the%20Salsa20%20nonce.pdf. Accessed 15 June 2022
11. Bernstein, D.J.: The Poly1305-AES message-authentication code. In: Gilbert, H., Handschuh, H. (eds.) FSE 2005. LNCS, vol. 3557, pp. 32–49. Springer, Berlin (2005). https://doi.org/10.1007/11502760_3, ISBN 978-3-540-26541-2
12. Friedman, J.H.: Greedy function approximation: a gradient boosting machine. Ann. Stat. JSTOR **29**, 1189–1232 (2001)
13. Chen, T., Guestrin, C.: XGBoost: a scalable tree boosting system. In: Proceedings of the 22nd ACM SIGKDD International Conference on Knowledge Discovery and Data Mining (KDD 2016), pp. 785–794. Association for Computing Machinery, New York, NY, USA (2016). https://doi.org/10.1145/2939672.2939785

A Property of a Quasigroup Based Code for Error Detection

Natasha Ilievska[✉]

Faculty of Computer Science and Engineering, Ss. Cyril and Methodius University, Skopje, Republic of Macedonia
natasa.ilievska@finki.ukim.mk

Abstract. Due to the importance of the accuracy of the transmitted data, codes for error-control play important role in the digital communications. While codes for error correction are more robust and slower, the codes for error detection work faster and therefore are more suitable in fast networks with small probability of errors. In the recent years, few quasigroup based codes for error detection have been developed. In this paper is consider one such quasigroup based code for error detection that is defined previously. The probability that there will be errors that the code will not detect and the number of surely detected incorrectly transmitted bits are already obtained. Since the code has good error-detecting capabilities, it deserves to be further analyzed. Now, in this paper, using simulations, it will be shown that when a quasigroup from a given set of order 4 quasigroups is used for coding, the code has the property of always detecting errors in transmission in the codewords in which the number of bits that are not correctly transmitted is odd number.

Keywords: Code for error detection · Quasigroup · Error-detecting capability

1 Motivation

Error control is important part in every communication system. In the fast networks with small probability of incorrect transmission of a bit, codes for error detection are more suitable than codes for error correction. Namely, since codes for error detection are simpler and faster, in networks with small probability of errors it is more practical to use them instead of codes for error corrections. In case of incorrect transmission of a data, that part of data is simply retransmitted.

Over the years, several codes for error detection have been developed. Several quasigroup based codes have also been developed over the last two decades. In this paper is considered a previously defined quasigroup based code for error detection [1]. In the previous work [1–6] is determined the power of the code to detect errors. First, it was concluded that the power of the code to detect errors depends on the quasigroup that is used for coding [1]. It was shown that there are 7 sets of order 4 quasigroups suitable to be used with this code. The

K. Zdravkova and L. Basnarkov (Eds.): ICT Innovations 2022, CCIS 1740, pp. 26–37, 2022.
https://doi.org/10.1007/978-3-031-22792-9_3

quasigroups in a same set have equal probability of undetected errors. In the papers [1, 2] are obtained general formulas for the probabilities that there will be errors that the code will not detect as functions of the input block length and the probability of incorrect transmission of a bit in the binary symmetric channel (BSC) in the case when arbitrary quasigroup is used for coding. Using this formula is obtained the corresponding probability when for coding is used a quasigroup from each of the 7 sets of order 4 quasigroups. The probability that there will be errors that the code will not detect is smallest when for coding are used the quasigroup from Set 1 in [2], therefore they are best for using with this code. All quasigroups in this set are so-called linear quasigroups.

In order to see where this code stands, the probability that there will be errors in transmission that the code will not detect when arbitrary quasigroup from the mentioned above Set 1 is used with this code is compared with the corresponding probability of few Cyclic Redundancy Check (CRC) codes as codes which are standard in error detection. CRC is defined six decades ago [7] and is based on the idea of seeing the message as a polynomial. This polynomial (the message) is divided by polynomial. The remainder from this division is the redundancy which is added to the input message. In the years back to the present day different polynomials are defined for coding with CRC code [8–10]. The code based on quasigroups has rate $1/2$, while the CRC is code with fixed length redundancy. It is reasonable to compare codes with equal rates. Therefore, the comparison is done when the codes have equal rates and lengths of the blocks are equal. Since the comparison showed that above-mentioned probability is smaller when code based on quasigroups is used, but also due to the fast coding and checking procedure, the research of this code has been continued. When working with linear quasigroups, of special interest are those quasigroups for which the constant term in the linear representation is the zero matrix. Hence, in [4, 5] is obtained the number of surely detected incorrectly transmitted bits when such a quasigroup from Set 1 is used for coding. In the previous work is shown that the difference in the probabilities that there will be errors that the code will not detect in the case when on one site quasigroups from Set 1 are used and on the other side the quasigroups in Set 2 from [2] are used is inconsiderable small. Therefore, the quasigroups from Set 2 are as good for coding as those from Set 1 in the light of the probability that there will be errors that the code will not detect. This is the reason why in the further examinations these quasigroups are investigated, too. All quasigroups in Set 2 are also linear quasigroups. The number of surely detected incorrectly transmitted bits when the coding is done with a qausigroup from Set 2 for which the constant term is zero matrix is obtained in [6]. In the same paper is concluded that the quasigroups from Set 2 surely detect equal number of bits that are not correctly transmitted as those from Set 1 when the length of the input block is greater than 2 symbols from a quasigroup of order 4. Therefore, since in this case the quasigroups from Set 2 have equal error-detecting capability as those in Set 1, the investigation of the code in this paper is also focused on the case when for coding is used a quasigroup from Set 2 for which the constant term is the zero matrix. Using simulations, it will be shown that the code has one interesting property when for coding is used such quasigroup - to detect the transmission errors always when the number of bits that are not correctly transmitted is odd number (as when for coding is used the order 4 quasigroup in [5]).

The organization of the paper is as follows. The definition of the code is given in Sect. 2. Also, in this section is given the set of quasigroups that are used in this paper, i.e., the quasigroups from Set 2 for which the constant term is the zero matrix. In Sect. 3 is given a brief description of the simulation process and the simulation results when for coding is used each of the order 4 quasigroups from the Set 2 for which the constant term is a zero matrix. At the end, the paper is concluded.

2   Definition of the Code

At the beginning of this section there is a brief explanation of the algebraic structures quasigroup and linear quasigroup.

A set Q together with a binary operation $*$ defined on it, that satisfy the condition that for every $u, v \in Q$ the equations

$$x * u = v \ \& \ u * y = v \tag{1}$$

have unique solution in Q along x and y is called a quasigroup. When Q is a finite set, the number of elements in it is called order of a quasigroup.

The code considered in this paper is analyzed when for coding are used linear quasigroups. A quasigroup $(Q, *)$ of order 2^q is said to be linear iff there exist invertible binary matrices A and B of order $q \times q$ and a binary matrix C of order $1 \times q$, such that for every two elements x and y from the quasigroup holds the following expression

$$x * y = z \Leftrightarrow \mathbf{z} = \mathbf{x}A + \mathbf{y}B + C \tag{2}$$

The bolded symbols in (2) are the binary representations of the corresponding symbol and $+$ is a binary addition. Therefore, linear quasigroup operations are in fact straightforward operations with binary matrices. As already stated, all quasigroups in the previously mentioned Set 1 and Set 2 are linear.

The code that is considered in this paper is defined in [1] as follows. Let the quasigroup $(Q, *)$ is used for coding and the symbols from the input messages are elements of Q. Each input message is divided in input blocks of length n symbols from Q and for each input block $a_0 a_1 \ldots a_{n-1}$ ($a_i \in Q, \forall i \in \{0, 1, ..., n-1\}$) the redundancy $d_0 d_1 \ldots d_{n-1}$, i.e., the redundant symbols $d_i, i \in \{0, 1, ..., n-1\}$ are calculated as follows:

$$d_i = a_i * a_{i+1 \ (mod \ n)}, i \in \{0, 1, \ldots, n-1\} \tag{3}$$

The binary form of the coded block $a_0 a_1 \ldots a_{n-1} d_0 d_1 \ldots d_{n-1}$, i.e., codeword, is transmitted through the BSC. Under the influence of the channel noises, some of the symbols may not be correctly transmitted. If the output block is $a_0'a_1' \ldots a_{n-1}'d_0'd_1' \ldots d_{n-1}'$, to check whether the block is correctly transmitted, the receiver checks whether

$$d_i' = a_i' * a_{i+1 \,(mod\, n)}', i \in \{0, 1, \ldots, n-1\} \qquad (4)$$

If for some $i \in \{0, 1, \ldots, n-1\}$ (4) is not satisfied, the receiver detects the error in transmission of the codeword and requires retransmission of the codeword. Otherwise, it concludes that the received block does not have errors.

But, it is possible to have a situation in which, in addition to the information characters, some redundant characters are incorrectly transmitted, too. For this reason, it is possible (4) to be satisfied for all $i \in \{0, 1, ..., n-1\}$ although the block is not correctly transmitted. Therefore, it is possible to have undetected incorrect transmission. Hence, for each code for error detection, its capability to detect errors is of key importance. It is defined by the probability that there will be errors that the code will not detect and the number of surely detected incorrectly transmitted bits. As already explained in the previous section, these parameters are already obtained. In this paper is discovered one more property regarding the error-detecting capability of the code when a quasigroup from the Set 2 from [2] for which the matric C in (2) is a zero matrix is used for coding. This set contains the quasigroups with lexicographical numbers 43, 93, 101 and 133, given in Fig. 1 and denoted with Q_1, Q_2, Q_3 and Q_4 respectively. The pairs of non-singular binary matrices A_1 and B_1, A_2 and B_2, A_3 and B_3 and A_4 and B_4, that represent these quasigroups (i.e., the matrices for which (2) is satisfied) are given in Fig. 2. When $C = [0\ 0]$, these pairs of matrices define the quasigroups Q_1, Q_2, Q_3 and Q_4, respectively.

In the simulation process described in the next section are used the quasigroups given in Fig. 1, i.e., Fig. 2.

$*$	0 1 2 3		$*$	0 1 2 3		$*$	0 1 2 3		$*$	0 1 2 3
0	0 1 3 2		0	0 2 3 1		0	0 3 1 2		0	0 3 2 1
1	3 2 0 1,		1	3 1 0 2,		1	1 2 0 3,		1	2 1 0 3
2	1 0 2 3		2	2 0 1 3		2	3 0 2 1		2	3 0 1 2
3	2 3 1 0		3	1 3 2 0		3	2 1 3 0		3	1 2 3 0
	Q_1			Q_2			Q_3			Q_4

Fig. 1. The order 4 quasigroups in Set 2 for which the matrix C in (2) is zero matrix

$$A_1 = \begin{bmatrix} 0 & 1 \\ 1 & 1 \end{bmatrix}, B_1 = \begin{bmatrix} 1 & 1 \\ 0 & 1 \end{bmatrix} \qquad A_2 = \begin{bmatrix} 1 & 0 \\ 1 & 1 \end{bmatrix}, B_2 = \begin{bmatrix} 1 & 1 \\ 1 & 0 \end{bmatrix}$$

$$A_3 = \begin{bmatrix} 1 & 1 \\ 0 & 1 \end{bmatrix}, B_3 = \begin{bmatrix} 0 & 1 \\ 1 & 1 \end{bmatrix} \qquad A_4 = \begin{bmatrix} 1 & 1 \\ 1 & 0 \end{bmatrix}, B_4 = \begin{bmatrix} 1 & 0 \\ 1 & 1 \end{bmatrix}$$

Fig. 2. Matrix representation of the quasigroups from Fig. 1

Example 1. This example demonstrates the coding and checking process for one input message. Let suppose that for coding is used a quasigroup Q_1 from Fig. 1. Let the input message be 02012321 and the input blocks have length 4 symbols from Q_1. Then, the input message is separated in blocks with length 4, i.e., the input message consists of two input blocks: 0201 and 2321. Each of them is coded.

For the first block 0201, the information characters are $a_0^1 = 0, a_1^1 = 2, a_2^1 = 0, a_3^1 = 1$. The redundant characters, calculated using (3) are:

$$
\begin{aligned}
d_0^1 &= a_0^1 * a_1^1 = 0 * 2 = 3 \\
d_1^1 &= a_1^1 * a_2^1 = 2 * 0 = 1 \\
d_2^1 &= a_2^1 * a_3^1 = 0 * 1 = 1 \\
d_3^1 &= a_3^1 * a_0^1 = 1 * 0 = 3
\end{aligned}
\tag{5}
$$

The coded input block is $a_0^1 a_1^1 a_2^1 a_3^1 d_0^1 d_1^1 d_2^1 d_3^1 = 02013113$. For the second block 2321, the information characters are $a_0^2 = 2, a_1^2 = 3, a_2^2 = 2, a_3^2 = 1$. The redundant characters, are:

$$
\begin{aligned}
d_0^2 &= a_0^2 * a_1^2 = 2 * 3 = 3 \\
d_1^2 &= a_1^2 * a_2^2 = 3 * 2 = 1 \\
d_2^2 &= a_2^2 * a_3^2 = 2 * 1 = 0 \\
d_3^2 &= a_3^2 * a_0^2 = 1 * 2 = 0
\end{aligned}
\tag{6}
$$

The coded input block is $a_0^2 a_1^2 a_2^2 a_3^2 d_0^2 d_1^2 d_2^2 d_3^2 = 23213100$. Now, the coded message, is $a_0^1 a_1^1 a_2^1 a_3^1 d_0^1 d_1^1 d_2^1 d_3^1 a_0^2 a_1^2 a_2^2 a_3^2 d_0^2 d_1^2 d_2^2 d_3^2 = 0201311323213100$. The message turned into binary form is 0010000111010111011100111010000. This is the message which is sent trough the BSC.

Let suppose now that the fourth, the seventeenth, twenty-sixth, thirty-first and thirty-second bits are not correctly transmitted, i.e., the receiver receives the message 0011000111010111001110011010011. This output message over the alphabet Q_1 is $a_0^{1'} a_1^{1'} a_2^{1'} a_3^{1'} d_0^{1'} d_1^{1'} d_2^{1'} d_3^{1'} a_0^{2'} a_1^{2'} a_2^{2'} a_3^{2'} d_0^{2'} d_1^{2'} d_2^{2'} d_3^{2'} = 0301311303212103$. The first output block in this message is $a_0^{1'} a_1^{1'} a_2^{1'} a_3^{1'} d_0^{1'} d_1^{1'} d_2^{1'} d_3^{1'} = 03013113$, while the second output block is $a_0^{2'} a_1^{2'} a_2^{2'} a_3^{2'} d_0^{2'} d_1^{2'} d_2^{2'} d_3^{2'} = 03212103$. As may be noted, the second information character from the first block, the first information character from the second block, but also the first and the last redundant characters from the second block are incorrectly transmitted.

To check if the first codeword is correctly transmitted, the receiver should check if the characters $d_0^{1'}, d_1^{1'}, d_2^{1'}$ and $d_3^{1'}$ are redundant characters for the block $a_0^{1'} a_1^{1'} a_2^{1'} a_3^{1'}$ according to (3), i.e., to check whether (4) is satisfied:

$$
a_0^{1'} * a_1^{1'} = 0 * 3 = 2 \neq 3 = d_0^{1'}
\tag{7}
$$

Since (4) is not satisfied when $i = 0$, the code detects the error in transmission of the first codeword and the receiver requires retransmission of that codeword.

Now, the receiver checks whether the second codeword is correctly transmitted:

$$
\begin{aligned}
a_0^{2'} * a_1^{2'} &= 0 * 3 = 2 = d_0^{2'} \\
a_1^{2'} * a_2^{2'} &= 3 * 2 = 1 = d_1^{2'} \\
a_2^{2'} * a_3^{2'} &= 2 * 1 = 0 = d_2^{2'} \\
a_3^{2'} * a_0^{2'} &= 1 * 0 = 3 = d_3^{2'}
\end{aligned}
\tag{8}
$$

In this case, since (4) is satisfied for all $i \in \{0, 1, 2, 3\}$, the errors are not detected (although they exist) and the block is accepted as a block without errors.

Example 2. Since the quasigroup used for coding in Example 1 is linear, coding and checking can be done using the binary matrices A_1 and B_1. In this case first the input blocks are turned into binary form, and then the redundant characters are calculated. For example, for the first input block 0201, the binary forms of the information symbols are $a_0^1 = [0\ 0]$, $a_1^1 = [1\ 0]$, $a_2^1 = [0\ 0]$ and $a_3^1 = [0\ 1]$. The redundant symbols are calculated using (2) and (3):

$$
d_0^1 = a_0^1 A_1 + a_1^1 B_1 = [0\ 0] \begin{bmatrix} 0 & 1 \\ 1 & 1 \end{bmatrix} + [1\ 0] \begin{bmatrix} 1 & 1 \\ 0 & 1 \end{bmatrix} = [1\ 1]
$$

$$
d_1^1 = a_1^1 A_1 + a_2^1 B_1 = [1\ 0] \begin{bmatrix} 0 & 1 \\ 1 & 1 \end{bmatrix} + [0\ 0] \begin{bmatrix} 1 & 1 \\ 0 & 1 \end{bmatrix} = [0\ 1]
$$

$$
d_2^1 = a_2^1 A_1 + a_3^1 B_1 = [0\ 0] \begin{bmatrix} 0 & 1 \\ 1 & 1 \end{bmatrix} + [0\ 1] \begin{bmatrix} 1 & 1 \\ 0 & 1 \end{bmatrix} = [0\ 1]
$$

$$
d_3^1 = a_3^1 A_1 + a_0^1 B_1 = [0\ 1] \begin{bmatrix} 0 & 1 \\ 1 & 1 \end{bmatrix} + [0\ 0] \begin{bmatrix} 1 & 1 \\ 0 & 1 \end{bmatrix} = [1\ 1]
$$

As can be seen, the same redundant symbols as in Example 1 are obtained, but now directly in binary form. The coded block is $a_0^1 a_1^1 a_2^1 a_3^1 d_0^1 d_1^1 d_2^1 d_3^1 = 0010000$ 111010111. The coding of the other input block continues in a same manner. If during the transmission the same errors occurs as in the previous example, i.e., if the first coded block is transmitted as $a_0^{1'} a_1^{1'} a_2^{1'} a_3^{1'} d_0^{1'} d_1^{1'} d_2^{1'} d_3^{1'} = $ 0011000111010111, then since

$$
a_0^{1'} A_1 + a_1^{1'} B_1 = [0\ 0] \begin{bmatrix} 0 & 1 \\ 1 & 1 \end{bmatrix} + [1\ 1] \begin{bmatrix} 1 & 1 \\ 0 & 1 \end{bmatrix} = [1\ 0] \neq [1\ 1] = d_0^{1'}
$$

the code detects the error in transmission.

3 Results

In order to experimentally show that the code detects the errors always when the number of incorrectly transmitted bits is odd number, the simulation procedure described in [5] is used. In short, for a given quasigroup Q and length of the codewords n bits, the simulation procedure consists of the following:

1) An input message with a length of several million symbols from Q is generated;
2) This message is divided in blocks with length $n/4$ symbols;
3) The blocks are coded using (3) and then turned into binary form are transmitted through a simulated BSC in which the probability of incorrectly transmitted bit is p;
4) Finally, for each $i \in \{1, 2, ..., n\}$ is counted the number of codewords in which i bits are not correctly transmitted ($bg[i]$) and for each such codeword, using (4) is checked whether the errors are detected, with which is obtained the number of codewords in which i bits are not correctly transmitted and the error is not detected ($bgno[i]$).

In the tables below are given the results from the simulation when for coding is used each of the four quasigroups from Fig. 1. In these tables, n is the length of the codewords in binary form. According to the definition of the code, if the input block has length k symbols from the order 4 quasigroup, then the coded block, i.e., the codeword will have length $2k$ symbols from the quasigroup. Each symbol from such quasigroup has 2 bits in the binary form, which implies that the binary form of the codeword will have length $4k$ bits. Therefore, the length of the codewords is a multiple of 4.

In order to obtain accurate results, the numbers $bg[i]$ of codewords in which i bits are not correctly transmitted must be large numbers for each $i \in \{1, 2, ..., n\}$. For each value of n, the probabilities p of incorrect transmission of a bit in BSC are chosen in a way this goal to be achieved.

Since when n increases, the probability that there will be errors that the code will not detect decreases [1,2], for larger values of n there are small number of codewords with undetected errors ($bgno[i]$), although the number of codewords which are not correctly transmitted ($bg[i]$) is large for every i. To obtain large number of codewords which are not correctly transmitted, for larger values of n (i.e., $n = 36$, $n = 40$ and $n = 44$), the generated input message is much longer, i.e., it is around 20000000 symbols from the quasigroup. For the other values of n, the generated message has length of around 2000000 symbols from the quasigroup. In case when $n = 44$, due to the extremely small probability that the errors will not be detected when more than 10 bit are incorrectly transmitted, the number of codewords in which the error is not detected is zero when more than 10 bits are not correctly transmitted ($bgno[i] = 0$ for $i > 10$).

From the simulation results presented in tables given below can be seen that all elements in the rows in which are the numbers of codewords in which odd number of bits are not correctly transmitted and the error is not detected, i.e., rows for $bgno[i]$ where i is odd number, are zeros. This means that when for coding is used a quasigroup from Fig. 1 then regardless of the length of the codewords, the code surely detects the transmission errors always when the number of bits that are not correctly transmitted is odd number (Tables 1, 2, 3 and 4).

Table 1. The number of codewords in which i bits are not correctly transmitted $bg[i]$ and the number of these codewords in which i bits are not correctly transmitted and the error in transmission is not detected $bgno[i]$ when the length of the codewords is n bits. For coding is used the quasigroup Q_1 from Fig. 1.

n	8	12	16	20	24	28	32	36	40	44
$bg[1]$	311303	242837	189635	147520	94110	68066	48509	34769	432306	337751
bgno[1]	0	0	0	0	0	0	0	0	0	0
$bg[2]$	69567	85176	90984	89667	93660	79812	66070	53364	542128	467933
$bgno[2]$	4945	0	0	0	0	0	0	0	0	0
$bg[3]$	218537	35703	26997	34638	60150	59392	56680	52057	432773	412845
bgno[3]	0	0	0	0	0	0	0	0	0	0
$bg[4]$	273340	80596	14007	9437	27265	44617	28009	36719	251950	264377
$bgno[4]$	35094	2981	132	36	55	63	29	22	90	191
$bg[5]$	218768	128823	33286	5988	39643	53166	39824	135391	173985	134323
bgno[5]	0	0	0	0	0	0	0	0	0	0
$bg[6]$	109724	150436	61247	14779	53030	50862	44200	175157	255739	55954
$bgno[6]$	15575	3885	418	37	29	13	3	3	4	2
$bg[7]$	198175	151375	87498	29522	58623	39691	40507	185368	307696	237663
bgno[7]	0	0	0	0	0	0	0	0	0	0
$bg[8]$	168298	142334	97854	48080	53223	25873	31344	165320	313278	271177
$bgno[8]$	0	6027	797	69	25	4	3	3	3	4
$bg[9]$		156862	87502	64058	40620	31786	20661	138052	283632	268943
bgno[9]		0	0	0	0	0	0	0	0	0
$bg[10]$		189372	60881	70549	26113	40225	37240	157782	289437	247078
$bgno[10]$		0	530	148	16	5	2	2	2	1
$bg[11]$		137553	32981	64020	20533	43670	31774	158478	260045	250931
bgno[11]		0	0	0	0	0	0	0	0	0
$bg[12]$		45620	13936	47952	32568	40863	23503	139524	206144	226919
$bgno[12]$		0	99	98	13	3	1	2	3	0
$bg[13]$			123098	29741	45739	33605	35601	137249	147803	237777
bgno[13]			0	0	0	0	0	0	0	0
$bg[14]$			105336	14748	53478	23911	27430	148567	215903	222598
$bgno[14]$			0	31	26	5	1	0	2	0
$bg[15]$			56325	69662	53644	33472	32724	144989	247232	188653
bgno[15]			0	0	0	0	0	0	0	0
$bg[16]$			14064	87520	45674	41379	34554	125951	254878	147382
$bgno[16]$			0	0	40	10	0	1	1	0
$bg[17]$				81979	33233	43252	32874	98400	237949	103169
bgno[17]				0	0	0	0	0	0	0
$bg[18]$				54898	52171	40370	27368	69753	201215	66597
$bgno[18]$				0	18	13	0	1	1	0
$bg[19]$				22974	65070	31656	35312	44700	156020	40054
bgno[19]				0	0	0	0	0	0	0
$bg[20]$				22974	65070	31656	35312	44700	109485	22318
$bgno[20]$				0	0	0	0	0	0	0

Table 2. The number of codewords in which i bits are not correctly transmitted $bg[i]$ and the number of these codewords in which i bits are not correctly transmitted and the error in transmission is not detected $bgno[i]$ when the length of the codewords is n bits. For coding is used the quasigroup Q_2 from Fig. 1.

n	8	12	16	20	24	28	32	36	40	44
$bg[1]$	311265	243107	190000	147629	93701	67416	48464	34500	432513	337425
bgno[1]	0	0	0	0	0	0	0	0	0	0
$bg[2]$	69139	85291	91172	89865	94624	79584	65855	53451	541520	467727
$bgno[2]$	4955	0	0	0	0	0	0	0	0	0
$bg[3]$	219086	35727	26860	34720	59847	59533	56770	52022	433082	412977
bgno[3]	0	0	0	0	0	0	0	0	0	0
$bg[4]$	273107	80550	13686	9334	22986	44415	28235	83104	93272	265486
$bgno[4]$	35287	2933	147	30	46	58	23	54	26	68
$bg[5]$	218249	129162	33306	5852	39371	53367	39773	135545	173754	134233
bgno[5]	0	0	0	0	0	0	0	0	0	0
$bg[6]$	109498	150737	60436	14905	53095	50819	43977	174882	255244	56228
$bgno[6]$	15554	3961	426	19	27	7	5	4	5	4
$bg[7]$	198265	151332	87653	29558	58548	39889	40403	185575	308592	237180
bgno[7]	0	0	0	0	0	0	0	0	0	0
$bg[8]$	167412	142364	98031	48126	53590	26156	31271	165552	242126	271925
$bgno[8]$	0	5926	803	90	2	4	2	5	2	2
$bg[9]$		157217	87844	64058	40398	31782	20762	137219	283890	268244
bgno[9]		0	0	0	0	0	0	0	0	0
$bg[10]$		189372	61189	70549	26056	40567	36877	158488	289369	247099
$bgno[10]$		0	485	141	12	5	2	2	0	0
$bg[11]$		137553	33350	64020	20269	43246	31542	158296	259745	251962
bgno[11]		0	0	0	0	0	0	0	0	0
$bg[12]$		45733	13975	47951	33017	41032	23577	112944	117969	226602
$bgno[12]$		0	104	104	12	2	1	1	1	0
$bg[13]$			123050	29332	45409	33418	35371	137257	169514	237158
bgno[13]			0	0	0	0	0	0	0	0
$bg[14]$			105827	14643	53826	23823	27479	149268	215607	222333
$bgno[14]$			0	37	13	3	1	1	1	0
$bg[15]$			56317	69587	53879	33525	32425	143882	246677	188883
bgno[15]			0	0	0	0	0	0	0	0
$bg[16]$			14168	87325	45304	40955	34933	126126	254702	146329
$bgno[16]$			0	0	21	4	2	2	0	0
$bg[17]$				82417	33453	43766	33180	98532	237425	103253
bgno[17]				0	0	0	0	0	0	0
$bg[18]$				55006	51503	40264	27395	69004	201408	67379
$bgno[18]$				0	9	10	1	2	1	0
$bg[19]$				23013	65332	31911	35275	44129	155683	40357
bgno[19]				0	0	0	0	0	0	0
$bg[20]$				4563	65237	21425	34587	25780	109793	22371
$bgno[20]$				0	0	1	0	0	1	0

Table 3. The number of codewords in which i bits are not correctly transmitted $bg[i]$ and the number of these codewords in which i bits are not correctly transmitted and the error in transmission is not detected $bgno[i]$ when the length of the codewords is n bits. For coding is used the quasigroup Q_3 from Fig. 1.

n	8	12	16	20	24	28	32	36	40	44
$bg[1]$	311211	242787	189461	147859	93812	67922	48655	34897	432992	336940
bgno[1]	0	0	0	0	0	0	0	0	0	0
$bg[2]$	69471	85520	91021	89865	94210	79385	66088	53339	541199	468711
$bgno[2]$	4930	0	0	0	0	0	0	0	0	0
$bg[3]$	218846	35853	27213	34433	59709	59898	56448	52276	432797	412993
bgno[3]	0	0	0	0	0	0	0	0	0	0
$bg[4]$	273930	80572	13885	9274	22777	45003	28315	84011	250833	265029
$bgno[4]$	35209	3037	130	35	57	60	28	64	71	70
$bg[5]$	218509	128959	33469	5931	39512	53257	39446	135417	173303	134302
bgno[5]	0	0	0	0	0	0	0	0	0	0
$bg[6]$	108975	151042	61153	14793	53554	50249	44223	174953	255727	56050
$bgno[6]$	15661	3964	393	28	22	9	4	5	2	4
$bg[7]$	198081	151575	87970	29541	58400	39661	40614	185268	308197	238449
bgno[7]	0	0	0	0	0	0	0	0	0	0
$bg[8]$	167854	141584	98396	47812	53165	26128	31157	104006	313610	272151
$bgno[8]$	0	6038	809	77	17	2	1	2	3	0
$bg[9]$		156988	86905	64256	40591	31778	20715	137691	284295	267715
bgno[9]		0	0	0	0	0	0	0	0	0
$bg[10]$		189577	60570	70578	26076	39914	36967	158364	210065	246603
$bgno[10]$		0	521	129	5	5	2	1	1	1
$bg[11]$		137474	33150	63917	20098	43628	31635	157917	259127	251323
bgno[11]		0	0	0	0	0	0	0	0	0
$bg[12]$		45503	13874	48014	33276	41094	23543	112799	207158	226740
$bgno[12]$		0	89	120	16	3	0	2	2	0
$bg[13]$			122848	29797	45454	33597	35482	137332	147176	238566
bgno[13]			0	0	0	0	0	0	0	0
$bg[14]$			105877	14845	53427	23877	27295	148782	216082	224085
$bgno[14]$			0	23	20	4	2	1	0	0
$bg[15]$			56223	69644	54043	33294	32896	144205	246941	189166
bgno[15]			0	0	0	0	0	0	0	0
$bg[16]$			13917	87317	45436	40988	34842	126202	254152	145542
$bgno[16]$			0	0	34	4	4	0	0	0
$bg[17]$				81990	33206	43650	32664	98571	237491	103032
bgno[17]				0	0	0	0	0	0	0
$bg[18]$				54594	51705	40258	27552	69538	201776	67085
$bgno[18]$				0	9	2	1	1	0	0
$bg[19]$				22911	65253	31864	35652	44595	155644	39849
bgno[19]				0	0	0	0	0	0	0
$bg[20]$				4673	65586	21579	34891	25439	109636	22296
$bgno[20]$				0	0	2	1	0	0	0

Table 4. The number of codewords in which i bits are not correctly transmitted $bg[i]$ and the number of these codewords in which i bits are not correctly transmitted and the error in transmission is not detected $bgno[i]$ when the length of the codewords is n bits. For coding is used the quasigroup Q_4 from Fig. 1.

n	8	12	16	20	24	28	32	36	40	44
$bg[1]$	311314	242658	189849	148473	93833	67526	48988	34475	432397	337321
bgno[1]	0	0	0	0	0	0	0	0	0	0
$bg[2]$	69176	85612	90557	89821	94353	79582	65897	53635	543268	467182
$bgno[2]$	4966	0	0	0	0	0	0	0	0	0
$bg[3]$	219444	35816	27057	34615	59709	59581	56288	52024	431398	413891
bgno[3]	0	0	0	0	0	0	0	0	0	0
$bg[4]$	273007	80564	13915	9282	22995	44479	28150	83861	93767	265275
$bgno[4]$	35143	2893	131	27	50	73	34	53	36	191
$bg[5]$	218377	129060	33423	5937	39093	53456	39807	135792	173511	134299
bgno[5]	0	0	0	0	0	0	0	0	0	0
$bg[6]$	109433	150171	60734	14955	53070	50725	44555	174974	254672	55934
$bgno[6]$	15741	3875	397	29	23	12	8	9	9	1
$bg[7]$	198049	150892	87269	29451	58738	39853	40216	185553	307617	237581
bgno[7]	0	0	0	0	0	0	0	0	0	0
$bg[8]$	167790	142310	97769	48334	53428	25849	31375	103194	241824	272303
$bgno[8]$	0	5970	763	76	26	2	1	1	2	4
$bg[9]$		157613	87309	63894	40656	31582	20709	138917	283752	268101
bgno[9]		0	0	0	0	0	0	0	0	0
$bg[10]$		188602	61302	69991	26431	40270	36727	158314	289206	247150
$bgno[10]$		0	511	146	11	7	1	2	2	1
$bg[11]$		137595	33649	64103	20162	43629	31581	157683	260128	252029
bgno[11]		0	0	0	0	0	0	0	0	0
$bg[12]$		45665	13945	48196	33004	41032	23302	139962	206787	226212
$bgno[12]$		0	104	97	21	4	1	2	0	0
$bg[13]$			123134	29632	45845	33259	35619	136693	168863	237521
bgno[13]			0	0	0	0	0	0	0	0
$bg[14]$			105644	14824	53557	23934	27617	149148	215369	222606
$bgno[14]$			0	32	28	5	0	0	0	0
$bg[15]$			56131	69794	53389	33612	32764	144358	246559	189210
bgno[15]			0	0	0	0	0	0	0	0
$bg[16]$			14142	87539	45731	41103	34993	125938	254107	145740
$bgno[16]$			0	0	24	3	1	0	0	0
$bg[17]$				82415	32778	43390	33025	98538	239052	103429
bgno[17]				0	0	0	0	0	0	0
$bg[18]$				54572	51642	40221	27399	69672	201821	67960
$bgno[18]$				0	8	5	0	0	1	0
$bg[19]$				23283	65727	31858	35347	44224	155419	40754
bgno[19]				0	0	0	0	0	0	0
$bg[20]$				4574	65580	21378	34707	25498	109903	22416
$bgno[20]$				0	0	3	0	0	0	0

4   Conclusion

In the focus of this paper is a quasigroup based code for error detection. The code is analyzed in the case when the order 4 quasigroups with lexicographical numbers 43, 93, 101 and 133 are used for coding. All these quasigroups are linear with constant term in their linear representation equal to a zero matrix. The simulations results suggest that when each of these quasigroups is used for coding, the code detects the transmission errors always when odd number of bits are not correctly transmitted.

Acknowledgements. This work was partially financed by the Faculty of Computer Science and Engineering at the "Ss.Cyril and Methodius" University.

References

1. Ilievska, N., Bakeva, V.: A model of error-detecting codes based on quasigroups of order 4. In: 6th International Conference for Informatics and Information Technology, Bitola, pp. 7–11 (2008)
2. Bakeva, V., Ilievska, N.: A probabilistic model of error-detecting codes based on quasigroups. Quasigroups Relat. Syst. **17**, 135–148 (2009)
3. Ilievska, N., Gligoroski, D.: Quasigroup redundancy check codes for safety-critical systems. In: 11th Advanced International Conference on Telecommunications IARIA-AICT, Brussels, pp. 72–77 (2015)
4. Ilievska, N.: Number of errors that the error-detecting code surely detects. In: Trajanov, D., Bakeva, V. (eds.) ICT Innovations 2017. CCIS, vol. 778, pp. 219–228. Springer, Cham (2017). https://doi.org/10.1007/978-3-319-67597-8_21
5. Ilievska, N.: Simulating the error-detecting capability of the error-detecting code. In: 41st International Convention on Information and Communication Technology, Electronics and Microelectronics (MIPRO), pp. 528–533. IEEE (2018). https://doi.org/10.23919/MIPRO.2018.8400091
6. Ilievska, N.: Towards an error-detecting code. In: Daimi, K., Alsadoon, A. (eds.) ICR 2022. AISC, vol. 1431, pp. 321–330. Springer, Cham (2022). https://doi.org/10.1007/978-3-031-14054-9_30
7. Peterson, W.W., Brown, D.T.: Cyclic codes for error detection. In: IRE 1961, vol. 49, no. 1, pp. 228–235. IEEE (1961). https://doi.org/10.1109/JRPROC.1961.287814
8. Koopman, P.: 32-bit cyclic redundancy codes for internet applications. In: The International Conference on Dependable Systems and Networks, pp. 459–468. IEEE (2002). https://doi.org/10.1109/DSN.2002.1028931
9. Koopman, P., Chakravarty, T.: Cyclic redundancy code (CRC) polynomial selection for embedded networks. In: International Conference on Dependable Systems and Networks, pp. 145–154. IEEE (2004). https://doi.org/10.1109/DSN.2004.1311885
10. Jones, D.T.: An Improved 64-Bit Cyclic Redundancy Check for Protein Sequences. University College London (2009)

Multi-access Edge Computing Smart Relocation Approach from an NFV Perspective

Cristina Bernad[1]([⊠])[iD], Vojdan Kjorveziroski[2][iD], Pedro Juan Roig[1][iD],
Salvador Alcaraz[1][iD], Katja Gilly[1][iD], and Sonja Filiposka[2][iD]

[1] Department of Computer Engineering, Miguel Hernández University (Elche),
Alicante, Spain
{cbernad,proig,salcaraz,katya}@umh.es
[2] Faculty of Computer Science and Engineering,
Ss. Cyril and Methodius University (Skopje), Skopje, North Macedonia
{vojdan.kjorveziroski,sonja.filiposka}@inki.ukim.mk

Abstract. This paper analyses the virtualised entities migration process implementation within the ETSI-compliant edge framework, considering the necessary multi-access edge computing (MEC) modules information interchange required for instantiation, termination and migration of MEC applications. Based on the variant of the MEC-NFV architecture and the functions of each element of it,, a communication process that includes network function virtualisation (NFV) interfaces is provided, as a step towards the unresolved challenge of modelling and developing a migration procedure that is aligned with the MEC standardisation process.

Keywords: Edge computing · Optimisation · Standardisation · Process design · Mobile nodes

1 Introduction

Edge computing is devised as a promising technology that will boost the development of latency sensitive and intensive computing mobile applications on localised premises instead of transmitting them to remote servers in the cloud. Extending virtualised compute and network resources from cloud infrastructures to edge microdatacentres closer to end users' mobile devices has the potential to enable highly dynamic service provisioning that will guarantee the performance of near real time heterogeneous services such as autonomous vehicles, industrial automation or extended reality. Effective integration of edge computing with a broad range of user equipment is expected to include Internet of Things (IoT) devices, unmanned aerial vehicle (UAVs), autonomous cars, etc. However, a lot more effort is needed in order to accomplish such expectations as edge computing requires the convergence of IT and telecommunications networking industry, and there are still many issues on the table for reaching an industry consensus.

K. Zdravkova and L. Basnarkov (Eds.): ICT Innovations 2022, CCIS 1740, pp. 38–48, 2022.
https://doi.org/10.1007/978-3-031-22792-9_4

Meanwhile, the academia started working towards new initiatives that open the door to providing intelligence to the edge infrastructure, directing its research efforts toward solutions that lead to an Artificial Intelligence (AI)-assisted edge [1]. Machine learning algorithms implemented at the edge of the network is the next step to complete the technological set that will guarantee proactive behaviour and the best possible performance in resource management optimisation at the edge layer.

Standardisation efforts are also being intensified during these last years from the European Telecommunications Standards Institute (ETSI) Multi-access Edge Computing (MEC) initiative Industry Specification Group (ISG), providing a MEC reference framework documentation [2]. With the advantages provided when coupling 5G with MEC, it becomes a necessity to align the 5G architecture with the MEC reference framework architecture so that their joint implementation can be done in a compatible, effective manner. 5G is promoting services such as telemedicine, smart cars, IoT or HD video that will be boosted when all 5G specifications such as high reliability (99.999%), high data rates (greater than 10Gbps) and ultra-low latency (less than 1ms) are resolved for heterogeneous devices and services. MEC FW provides a scenario where the computation and network resources are closer to the end user and thus it achieves a very low latency with the already available technology. Thus, the initial ETSI MEC reference architecture has recently been extended and aligned to the existing Network Function Virtualisation (NFV) interfaces that provide the basis of the 5G core implementation in order to achieve compliance with the reference documents provided by the ETSI ISG NFV group. Although the major steps towards implementing MEC with NFV have been defined, there are still a number of open issues that need to be addressed with the implementation of smart relocation using MEC application migration [3]. Effective implementation of smart relocation is essential for advanced MEC systems that are implemented in a highly mobile environment such as 5G, where the benefits of MEC such as low latency can be retained only if the MEC services continuously remain in the closest possible vicinity of the users. In this paper, we focus on the MEC-NFV reference framework to conceptually design the smart relocation migration problem following the latest ETSI guidelines. Smart relocation is defined as the necessity to migrate the MEC applications to adapt to the changing network performances for mobile end-users. Namely, as the user enters a new service area, the virtualised entities that provide MEC services to the user's mobile user equipment need to be moved to the closest MEC host in order to continue guaranteeing ultra-low latency.

The remainder of this paper is organised as follows. In Sect. 2 we review some of the latest contributions focusing on virtualised entities migration in MEC-NFV environments. Section 3 introduces the MEC-NFV reference architecture and focuses on the description of the communication processes required for instantiating, terminating and migrating virtualised entities. The paper ends with a discussion and conclusions Sect. 4.

2   Related Work

Efficient resource management has been a very popular research topic ever since the introduction of edge computing [4]. However, while lots of attention has been given to the efficient initial placement of an application on a particular edge host, the problem of efficient migration in mobile environments has received less scrutiny [5]. Lately, the work on edge computing use cases such as autonomic vehicles, where mobility is an intrinsic part of the scenario, has given a tremendous rise to the approaches that aim to solve the problem of smart relocation using optimised migration approaches [6]. A few examples include work focusing on optimising resource allocation and migration in multi-cell scenarios [7], delay and mobility-aware approaches based on probabilistic methods [8], as well as on mobile agents [9], and Markov decision processes [10]. When it comes to the implementation of these algorithms and strategies, the virtualisation infrastructure of MEC applications is mostly implemented using virtual machines. Lately, there are several studies that explore containers as means to implement MEC such as [11].

When considering NFV, there are some other initiatives that tackle the migration problem by taking advantage of the concept of Service Function Chains (SFC) that is used to propose optimisation algorithms based on Dijkstra for enabling seamless migration through SFC reconfigurations [12]. Another example is focusing on optimising resource allocation for VNF migrations using genetic algorithms [13]. Recently, there are a number of studies that use AI to resolve smart relocation problems such as [14] that uses deep reinforcement learning strategies and [15] that introduces cognitive edge computing.

Recognising that the 5G and MEC integration can be achieved more easily if NFV is used for both, implementations of this blended environment are also being studied [16]. However, it is important to remark that the previously mentioned contributions do not consider the NFV-based MEC generic architecture in their proposals and, therefore, their research is not specifically aligned to the "smart relocation function" defined in [17] that provides details defining the process of transferring an instance of a MEC application to maintain the quality of service for the users. And while there is literature that addresses the MEC-NFV implementation, very few papers discuss the implementation of smart relocation in this setting. In these examples, such as [18], the authors do not consider the existence of two separate orchestrators and the rest of the specific MEC components that are identified in the ETSI MEC-NFV reference architecture. Thus, these approaches can not be used as a standardised approach to implementing smart relocation and more work is needed to define this process.

3   MEC-NFV Management Flows

Having in mind that the underlying architecture that is used to build the core 5G components is based on NFV elements, the MEC integration into the 5G system can be done in a smoother fashion if the MEC components can be implemented

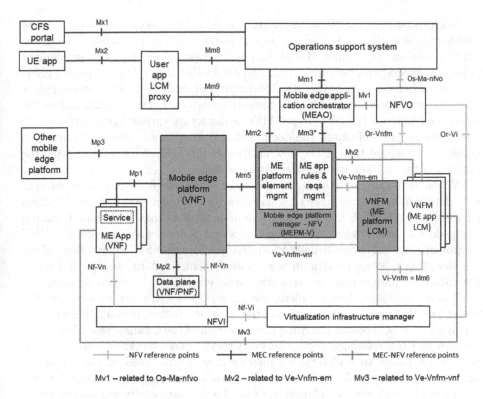

Fig. 1. NFV-based MEC generic architecture. (Color figure online)

using NFV entities [16]. For this reason, ETSI has been working for years on defining a framework and an architecture based on MEC-NFV [2].

The main idea of the MEC-NFV implementation is to implement MEC applications as Virtual Network Functions (VNFs), see Fig. 1. In this setting, there is a hierarchical setup of two separate orchestrators, the Mobile Edge Application Orchestrator (MEAO) that decides when and where to instantiate, migrate and terminate a MEC application, and the Network Function Virtualisation Orchestrator (NFVO) that takes over the responsibilities to create and manage the MEC applications as VNFs. The management of each individual MEC host, i.e. server hosting MEC applications, is done using the Mobile Edge Platform implemented as a separate VNF. Each set of co-located MEC hosts are managed by a virtualised implementation of the Mobile Edge Platform Manager (MEPM-V) that is in charge of specifying the MEC applications rules and requirements. The lifecycle management of each MEC application is delegated to the VNFM.

In an urban, dense 5G settings consider one or more base stations collocated with microdatacenters made out of several MEC hosts making up a mobile edge platform. Each service area has a MEPM-V that manages the grouped MEC hosts providing virtual MEC resources for that area and the orchestrators are used for a higher-level network-wide management.

Each element of the NFV-based MEC architecture uses a reference point to communicate with another entity as represented in Fig. 1. Initially, the ETSI MEC document [2] defined the MEC reference points (in blue colour) for the generic reference architecture, which can be divided into reference points referring to the functionality of the MEC platform (Mp), management reference points (Mm) and the reference points that connect to external entities (Mx). Later, the reference points for the NFV-based MEC architecture variant (green color) were also included and defined in the document and some more reference points (red colour) were added to join the new entities of the MEC-NFV variant with the entities of the general architecture (Mv2 and Mv3).

When analysing the migration process of VNFs, we find that there are specific details to be defined since they have not been subject to the standardisation carried out by ETSI MEC ISG yet. The main issue is that NFV based components do not include the explicit interfaces and processes for transparent VNF migration since this has not been a requirement for the NFV architecture. Thus, a series of open issues remain on how to implement the MEC application smart relocation for mobile end user scenarios in an efficient and uniform manner.

Aiming to tackle this problem, we have decided to start by defining the workflows that need to be implemented in order to instantiate, migrate and terminate a MEC application using the definition of the standardised MEC-NFV entities and the reference points between them. These workflows are intended to make it easier to understand which entities are involved in the migration process and what are their roles, as well as how all interfaces come together and where we should focus our attention to be able to successfully implement smart relocation.

3.1 MEC Application Instantiation

When we talk about a virtualised MEC-NFV environment, MEC applications are implemented as VNFs reached via user accessible MEC services. We detail out the steps taken during the instantiation process flow (see Fig. 2) that is activated when a new MEC application needs to be created.

1. A user, through an application that is running on his device (device app), sends a request to his service provider to instantiate a new MEC application via the Mx2 reference point, that connects with the user application lifecycle management (LCM) proxy which authorises requests from device application.
2. The user application LCM proxy exchanges, then, information with the Operation Support System (OSS) and the Multi-access edge orchestrator (MEAO) for the management of this instantiation request via the Mm8 in the MEC system.
3. The OSS decides whether to grant the request based on the user subscription information and service provider policies. OSS forwards approved requests to the MEAO for further processing through the Mm1 reference point.

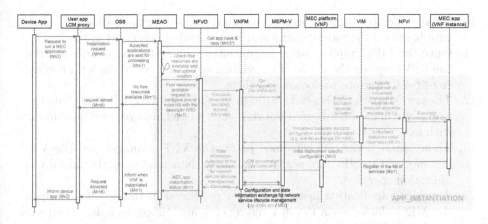

Fig. 2. MEC application instantiation process flowchart.

4. The MEAO communicates with the MEC Platform manager-NFV (MEPM-V) via Mm3 and obtains the application rules, requirements and defined traffic rules, if any.

5. Since the MEAO is the centre of the MEC system at the management level, it must have a global view of the MEC system's free resources, available services and topology. The MEAO decides on the optimal location for the new MEC application based on this information. In case there are free resources, the MEAO requests the Network Function Virtualisation Orchestrator (NFVO) to configure one or more Network Services (NSs) to manage the VNFs of the MEC application, generating and on-boarding an Network Service Descriptor (NSD), and requesting the instantiation of an NS. The descriptor is a template used by the NFVO with information for the instantiation of the NS, formed by one or more VNFs. In case there are no free resources, the MEAO informs the OSS through Mm1 which, in turn, will inform the User application LCM proxy through the Mm8 interface.

6. The next step is to reserve the necessary resources for deploying the requested MEC service and then, the allocation of the application is requested. The flow of this communication starts at the NFVO via Or-Vnfm to the Virtual Network Function Manager (VNFM), since the NFVO manages all VNF LCM operations with VNFM.

7. As the ME Platform VNF is considered as a network function, the VNFM gets VNF configuration of the MEPM-V through the Ve-Vnfm-em interface.

8. Likewise, the VNFM via Vi-Vnfm exchanges information with the Virtualisation Infrastructure Manager (VIM) about resource allocation. The VIM is responsible, among other functions, of allocating, managing and releasing virtualised resources of the Virtualisation infrastructure.

9. Therefore, the VIM performs the allocation of specific resources for the VNF in the Network Functions Virtualisation Infrastructure (NFVI), in response to the request to allocate resources through Nf-Vi interface.

10. Finally, the MEC application is instantiated on the virtualisation infrastructure based on the requirements and configuration described in the NSD.
11. Once the MEC application is instantiated, the system is informed. First, the NFVI reports the status of the virtualised resources to the VIM through the Nf-Vi interface that connects them.
12. The next step, the VIM sends through Vi-Vnfm to the VNFM the configuration of virtualised hardware resources and the exchange of state information such as events.
13. Then, through the Mv3 reference point, the VNF Manager and the VNF instance of the MEC application exchange data related to the initial configuration specified.
14. The MEC application instance is then registered in the list of services to the MEC platform via the Mp1 reference point. Since, as it has been said previously, VNF instances act in virtualised environments as another service managed by the VNFM.
15. Through the Ve-Vnfm-vnf interface, the VNF manager coordinates with the MEC LCM platform to manage the subscription on the LCM event.
16. Next, configuration and status information is exchanged between the MEC-NFV platform manager and the VNFM for network service lifecycle management via two interfaces that are: Mv2 and Ve-Vnfm-em.
17. Then, the NFVO, as responsible for the management of the life cycle of the network services, exchanges information with the VNFM about the VNF status required for network service lifecycle management through O-Vnfm.
18. Through the Mv1 interface, the NFVO informs the MEAO about the ME app VNF mapping and state as the NFVO is in charge of the orchestration of the set of ME app VNFs as one or more NFV NSs.
19. Likewise, the MEAO informs the OSS through the Mm1 interface that the requested instance has been carried out.
20. Finally, the OSS informs the user application LCM proxy that the request has been accepted using the Mm8 interface and this, in turn, informs the Device application via Mx2.

3.2   Migration

In case the MEC application has to be migrated to guarantee Quality of Service (QoS) requirements such as low latency when changing service areas as the device is moving, a clone of the MEC application must be instantiated on a more optimal host with enough available resources. Therefore the information from the old VNF must be copied to the new instance. Once the transfer is complete, the old VNF must terminate. Figure 3 shows the flow of the migration process based on the trigger from the location service [2] which provides location-related information for authorised applications. This information is sent through the Mp1 interface to the MEC platform (VNF) where one of its functions is to host MEC services. Thus, the location of the MEC application will be updated.

In this way, the MEC platform is informed where are the users that use the hosted MEC services. This location information travels through the Mm5 interface

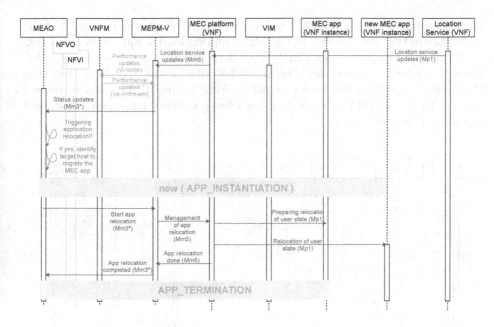

Fig. 3. Migration process flowchart of a MEC application.

from the MEC Platform (VNF) to the MEPM-V that manages the life cycle of the MEC applications. To ensure QoS, the system must keep track of the performances as the users interact with the MEC applications. For these purposes, the VIM sends performance information about the virtualised resources to the VNF Management via Vi-Vnfm. From there these performance updates are sent to the MEPM-V via Ve-Vnfm-em. The performance status together with the location information for all MEC applications are sent to the MEAO through the Mm3 interface. The MEAO needs to have this information since it must have an overview of the MEC system and must detect when it is necessary to migrate a MEC application when its QoS requirements are not met. In case the QoS requirements of a MEC application are not met, the MEAO must identify a new target host where the MEC application is going to be migrated. To implement the smart relocation feature the MEAO needs to start a new MEC application on the target host, initiate relocation and then terminate the old application instance on the original host.

Once an instance of the new MEC application has been created, the MEAO starts relocating the application context and user state in coordination with the MEPM-V via Mm3. At the same time, the current running application instance status must be maintained as this process must be transparent to the user. The MEPM-V with the information received from the MEAO, such as traffic rules, manages the relocation of the MEC application through the Mm5 interface together with the MEC Platform (VNF). Next, the MEC Platform (VNF) through Mp1 prepares the relocation of the user state. Once the user sessions have been transferred, the MEC Platform (VNF) performs the relocation of the user state to the new instance of the MEC application via Mp1. Next, the MEC platform (VNF) informs the MEPM-V that the relocation has been completed through the Mm5 interface. And the MEPM-V informs the MEAO via Mm3 that the relocation process has been completed. Finally, the MEAO must terminate the old MEC application. To do so it launches a termination process of the old instance of the application.

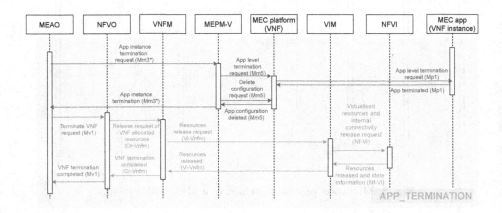

Fig. 4. Termination process flowchart of a MEC application.

3.3   Termination

The application instance termination process takes place in two steps (see Fig. 4): first, the ME platform configuration and data plane must be removed, and the second step is to terminate the ME app VNF. To do this, the MEAO communicates with the MEPM-V through Mm3 to request the termination of the application instance. The request arrives at the MEC platform (VNF) through the Mm5 interface. The MEC platform then sends an application level termination request to the old MEC application instance via Mp1. After the application has been terminated, the MEPM-V asks the MEC platform (VNF) to remove the platform configuration via Mm5. Once the confirmation is received, MEPM-V informs the MEAO that the application instance termination process has been completed by Mm3. The next step takes the MEAO to request the NFVO to terminate the VNF of the application via Mv1. To do this, the NFVO communicates with the VNFM by Or-Vnfm to request the release VNF allocated resources. The VNFM requests the release of these resources to VIM through Vi-Vnfm. Then the VIM, via Nf-Vi, requests the NFVI to release the visualised resources and internal connectivity. The NFVI releases the virtualised resources and sends status information back to the VIM, which forwards the confirmation that the resources have been released to the VNFM through Vi-Vnfm, that sends the confirmation that the termination of the VNF has been completed to the NFVO via `Or-Vnfm`. The NFVO by `Mv1` communicates it to the MEAO, which must be aware of the status of the whole MEC system.

4 Discussion and Conclusions

The emergence of new generation devices with ultra low latency requirements is increasing every day. There are many works that are committed to uniting MEC NFV-based and 5G technologies but there are still some concepts to be defined. With this work we want to provide a solution to the orchestration of the migration process in order to further the work done by ETSI and tackle the smart relocation issue in MEC-NFV implementations. The identification of the roles of each MEC component and the activation of the defined interfaces needed to implement smart relocation are a first step of the research. Our future work will focus on bridging the gap of implementing smart relocation in MEC-NFV by detailed analysis of the data exchanged via the identified interfaces and, finally, a practical implementation of the outlined workflows.

The ETSI has not yet defined the smart relocation process in MEC NFV-based environments, even though their documents indicate that the MEC system that supports smart relocation must be able to perform relocation of a MEC application instance from one MEC host to a different host within the system or to cloud system outside the MEC system. Therefore, in this work we assign to each of the involved components of the NFV-based MEC generic architecture a specific part in the smart relocation process. We produce a mapping of functions and blocks of the NFV-based MEC architecture, thus continuing the work addressed by the ETSI. As it can be seen in [2] this standardisation work continues with a new variant of the general MEC-NFV reference architecture extending it with federations. In a federated environment MEC services and applications are shared with each federation having its own computing resources at the edge. In this new variant of the MEC architecture a new element is added, MEC Federator (MEF), that allows the MEC system to communicate with other MEC systems and other non-MEC systems. In this constellation, the smart relocation function must allow the migration of the MEC application between MEC systems as well as between a MEC system and another non-MEC system, which will be addressed in future works.

References

1. Koufos, K., et al.: Trends in intelligent communication systems: review of standards, major research projects, and identification of research gaps. J. Sens. Actuator Netw. **10**, 60 (2021)
2. ETSI GS MEC 003: "Multi-Access Edge Computing (MEC); Framework and Reference Architecture". V3.1.1 (2020)
3. Bernad, C., Roig, P.J., Filiposka, S., Alcaraz, S., Gilly, K.: Challenges of implementing NFV-based multi-access edge computing environments. In: 2021 29th Telecommunications Forum (TELFOR), pp. 1–4 (2021)

4. Hong, C.H., Varghese, B.: Resource management in fog/edge computing: a survey on architectures, infrastructure, and algorithms. ACM Comput. Surv. (CSUR) **52**(5), 1–37 (2019)
5. Wang, S., Xu, J., Zhang, N., Liu, Y.: A survey on service migration in mobile edge computing. IEEE Access **6**, 23511–23528 (2018)
6. Jehangiri, A.I., et al.: Mobility-aware computational offloading in mobile edge networks: a survey. Cluster Comput. **24**(4), 2735–2756 (2021)
7. Liang, Z., Liu, Y., Lok, T.M., Huang, K.: Multi-cell mobile edge computing: joint service migration and resource allocation. IEEE Trans. Wirel. Commun. **20**, 5898–5912 (2021)
8. Xu, M., et al.: PDMA: probabilistic service migration approach for delay-aware and mobility-aware mobile edge computing. Softw. Pract. Exp. **52**(2), 394–414 (2022)
9. Guo, Y., Jiang, C., Wu, T.Y., Wang, A.: Mobile agent-based service migration in mobile edge computing. Int. J. Commun. Syst. **34**(3), e4699 (2021)
10. Wang, S., et al.: Dynamic service migration in mobile edge computing based on Markov decision process. IEEE/ACM Trans. Netw. **27**(3), 1272–1288 (2019)
11. Meng, X., Lu, W.: Container-based fast service migration method for mobile edge computing. J. Circuits Syst. Comput. 2250117 (2021)
12. Li, B., Cheng, B., Chen, J.: An efficient algorithm for service function chains reconfiguration in mobile edge cloud networks. In: Proceedings of the 2021 IEEE International Conference on Web Services, ICWS 2021, pp. 426–435 (2021)
13. Kiran, N., Liu, X., Wang, S., Yin, Ch.: Optimising resource allocation for virtual network functions in SDN/NFV-enabled MEC networks. IET Commun. **15**(13), 1710–1722 (2021)
14. Gao, Z., at al.: Deep reinforcement learning based service migration strategy for edge computing. In: 2019 IEEE International Conference on Service-Oriented System Engineering (SOSE), pp. 116–1165 (2019)
15. Chen, M., et al.: A dynamic service migration mechanism in edge cognitive computing. ACM Trans. Internet Technol. (TOIT) **19**(2), 1–15 (2019)
16. Blanco, B., et al.: Technology pillars in the architecture of future 5G mobile networks: NFV, MEC and SDN. Comput. Standards Interfaces **54**, 216–228 (2017)
17. ETSI GS MEC 017: "Deployment of Mobile Edge Computing in an NFV environment". V1.1.1 (2018)
18. Afrasiabi, S.N., et al.: Application components migration in NFV-based hybrid cloud/fog systems. In: 2019 IEEE International Symposium on Local and Metropolitan Area Networks (LANMAN), pp. 1–6 (2019)

Artificial Intelligence and Deep Learning

MACEDONIZER - The Macedonian Transformer Language Model

Jovana Dobreva[✉], Tashko Pavlov, Kostadin Mishev, Monika Simjanoska,
Stojancho Tudzarski, Dimitar Trajanov, and Ljupcho Kocarev

Faculty of Computer Science and Engineering, Ss. Cyril and Methodius University in
Skopje, Skopje, North Macedonia
{jovana.dobreva,kostadin.mishev,monika.simjanoska,
dimitar.trajanov,ljupcho.kocarev}@finki.ukim.mk,
{tashko.pavlov,stojancho.tudzharski}@students.finki.ukim.mk

Abstract. Contextualized language models are becoming omnipresent
in the field of Natural Language Processing (NLP). Their learning representation capabilities show dominant results in almost all downstream
NLP tasks. The main challenge that low-resource languages face is the
lack of language-specific language models since their pre-training process requires high-computing capabilities and rich resources of textual
data. This paper describes our efforts to pre-train the first contextual
language model in the Macedonian language (MACEDONIZER), pre-
trained on a 6.5 GB corpus of Macedonian texts crawled from public
web domains and Wikipedia. Next, we evaluate the pre-trained version
of the model on three different downstream tasks: Sentiment Analysis
(SA), Natural Language Inference (NLI) and Named Entity Recognition (NER). The evaluation results are compared to the cross-lingual
version of the RoBERTa model - XML-RoBERTa. The results show
that MACEDONIZER achieves state-of-the-art results in all downstream
tasks. Finally, the pre-trained version of the MACEDONIZER is made
for free usage and further task-specific fine-tuning via HuggingFace.

Keywords: MACEDONIZER · Contextualized language model ·
Macedonian · Sentiment analysis · Natural Language Inference ·
Named Entity Recognition · Pre-training

1 Introduction

Natural language-based contextualized models have become a significant subject
in the field of Natural Language Processing (NLP) due to their enormous power
to learn word and sentence representations that are not only grammatically correct, but also have semantic meaning. As such their application is wide, since the
initial product is a representation of general-purpose language model that can
then be further fine-tuned for a specific NLP task on smaller labeled dataset.
Thus, the generalized language model can be easily brought to an expert, specialized level in the field of finance, medicine, biology, and other complex areas at
which as a human expert it is almost impossible to cover that span of knowledge.

© The Author(s), under exclusive license to Springer Nature Switzerland AG 2022
K. Zdravkova and L. Basnarkov (Eds.): ICT Innovations 2022, CCIS 1740, pp. 51–62, 2022.
https://doi.org/10.1007/978-3-031-22792-9_5

The evolution of the architectures for text representations starts from methods for lexicon-based knowledge extraction, then statistical methods, word encoders, sentence encoders that have evolved to be even language-agnostic, up to the most popular era of NLP transformers. At its basic, a transformer architecture encompasses encoder and decoder to transform one sequence into another. The elimination of the recurrent layer upon which the previous sequence-to-sequence methods are based on, and introducing the self-attention mechanism [33] solves the three concepts that are needed to be fulfilled when building contextualized embeddings and those are: computational complexity, parallelization, and studying the sequence's long-term relationships between the terms [24]. As such, transformers have the ability to produce text-representations in an unsupervised manner, meaning the design of the transformer is extended to include an unlabeled text corpus upon which the objective function used by the transformer produces text representations. Although this task requires a lot of processing resources, transfer learning allows for the application of the learned token or general sentence representations in a variety of other tasks. Those representations are later fine-tuned to recognize the specifics of particular field's context. After the last hidden state, fine-tuning is done by adding another dense layer [11]. The fine-tuning is considered a supervised learning since the input dataset in this case is a labeled dataset, meaning the generalized model must be specialized in a particular field at which classification/regression tasks apply.

Even not a pioneer in the field, the transformers' "beauty" has been uncovered by introducing the revolutionary language representation model - BERT (Bidirectional Encoder Representations from Transformers) [11]. The novelty of BERT is in its ability to overcome the limitations of previous language models by building bidirectional masked language model from large unlabeled text corpus that predicts randomly masked terms in the sentence, enhancing the contextual information of the words. BERT comes in two versions, basic and large version, with a difference in the number of encoder layers (12 vs. 24), hidden size (768 vs. 1024), the number of multi-head attention heads (12 vs. 16), and number of parameters (110M vs. 340M). Both generalized models are trained on English Wikipedia and BookCorpus [36]. Versions of specialized BERT models are later published in various domains, such as FinBERT in the finance domain [3], Med-BERT in medicine [29], etc.

BERT has intrigued many initiatives that tried to address the disadvantages of BERT. XLNet [35] is one of the architectures that improved some parts of the architectural design and outperformed BERT in 20 different tasks. The improvements are in a context of introducing tokens dependencies instead of assuming independence as in the case of BERT. Then, XLNet applies Permutation Language Modeling (PLM) to maximize the anticipated log-likelihood of a sequence given all conceivable word permutations in a sentence in order to capture bidirectional context. XLM [14] goes even further to model cross-lingual features by using the following objectives: next token prediction (Causal Language Modeling), masking random tokens in a sentence (Masked Language Modeling - similar to BERT), and using simultaneous streams of text data in various languages (Translation Language Modeling). As previously mentioned, those tasks are computationally expensive, thus Google Research and Toyota Technological Institute provided a joint effort to release a smaller and more scalable successor of BERT - ALBERT [15]. The two-parameter reduction methods behind ALBERT to lower memory consumption and to increase training speed are cross-layer parameter sharing and sentence ordering objectives. As such ALBERT has managed to outperform BERT in several tasks, one of which is text classification [1]. An architecture with similar purpose to reduce the size of BERT is DistilBERT [30], which succeeded in decreasing the size of a BERT model by 40%, while retaining 97% of its language processing abilities and also to increase the training speed by 60%.

The evolution of the initiatives for reduction of BERT eventually led to the alternative optimized version of BERT introduced by Facebook research team - RoBERTa [18]. RoBERTa has been pre-trained on a dataset ten times larger and by removing the Next Sentence Prediction (NSP) objective, and adding a dynamic masking of words during the training process. Those changes have led to outperforming BERT in many NLP tasks, e.g., including the field in finance as presented in the comprehensive evaluation of all state-of-the-art models by 2020 [24]. Then follow similar trials for multilingual models, and as such is presented XLM-RoBERTa (XLM-R) model trained on 100 different languages with a use of 2.5TB of filtered CommonCrawl data [9].

Even though RoBERTa has shown outstanding performance compared to BERT, novel transformer-based architectures rapidly come to light, however, their demanding for resources increases with their complexity. BART [17] is one of the transformers that has shown to outperform all the known transformers up to 2019 on tasks in the field of finance [24] such as text summarizing and question answering. BART makes use of the benefits of both the GPT autoregressive (left to right dependence on the words in a sentence) decoder and the bidirectional encoder from BERT. The GPT-2 language model, which is the upgrade for GPT, has 1.5 billion parameters and was trained on 8 million web pages. The objective of GPT-2 is rather different and simple, that is to predict the next word, given all of the previous words within some text [27].

Newer architectures such as the model Text-to-Text Transfer Transformer (T5) has reached state-of-the-art performance on 18 out of 24 tasks considered [28], however, the number of parameters raised to 11 billion. GPT-3, an autoregressive language model, is even more extreme case trained with 175 billion parameters, which is ten times the number of parameters than any previous non-sparse language model [5]. The main idea is to achieve an avoidance of the need to collect large labeled training datasets for individual NLP tasks, by scaling up language models to improve task-agnostic few-shot performance.

Hereupon, the future must be focused on researching methods that will achieve stronger performance with cheaper models, or a distillation of large models is needed to a manageable size that will be suitable for real-world applications. In that direction is the presentation of the ELECTRA model at which the researchers suggest a new pre-training task named replaced token detection instead of the pre-training methods due to the masked language modeling that are ineffective in terms of computation as they use only a small fraction of tokens for learning [8]. Going back to the BERT-based models, DeBERTa (Decoding-enhanced BERT with disentangled attention) is the new model that enhances the BERT and RoBERTa models using two innovative methods: a disentangled attention mechanism, and an enhanced mask decoder. Compared to RoBERTa-Large, the DeBERTa model performs better even trained on half of the training data [13].

This paper proposes the first context-based framework for Macedonian language representation. To achieve a good trade-off between the demanding for resources and the performance of the model, we have used RoBERTa to create a generalized Macedonian language model.

The rest of the paper is organized as follows. Given the problem of training a model on language different from English, in the following Sect. 2 we present some of the efforts for other languages. The datasets used for Macedonian language are presented in Sect. 3, together with the RoBERTa training process as well as the dataset pre-processing done for this task. The downstream evaluation tasks are explained in Sect. 4 and the evaluation results are given in Sect. 5, together with a discussion on the models and the obtained results in Sect. 6. Finally, the conclusion of the research as well as the directions for future work are given in Sect. 7.

2   Related Work

The presented generalized language models have shown to perform very good for English language. However, when it comes to pre-training them for other language than English, the first constraint is the limitation of resources and the second is the creation of dataset that might not be available for some languages. It is quite simple to adapt models to languages that are genetically close to each other [34]. For example, both Dutch and English are WestGermanic languages. Also despite the fact that for example Italian is more distant Romance language from the same Indo-European language family, the order of the words at sentence level is still similar to English.

French is also part of the Romance language group, and CamemBERT [22] is a French language model for which the authors have shown that web scraped data is preferable to the use of Wikipedia articles data. Even more, relatively small dataset of only 4GB produced results as good as when used 130GB of data. Newer French model is FlauBERT [16] which slightly outperforms CamemBERT at some NLP tasks. Similar efforts are done for Brazilian (Portuguese) language - BERTimbau [31] and its competitor PTT5 [6].

For the languages that are distant from English, Arabic has a less developed syntax but is morphologically richer in terms of resources.

AraBERT [2] is a model for Arabic language whose training has unveiled some important conclusions, e.g., the data size is found to be important factor for the boost in performance. For the success of AraBERT 24GB of data have been used, whereas only 4.3G Wikipedia data has been used for the multilingual BERT. There is also difference in the vocab size used in the multilingual BERT (2k tokens) compared to AraBERT (64k tokens). The large data size brought more diversity in the pre-training distribution. Also the pre-segmentation applied before BERT tokenization improved the performance of the sentiment analysis and the question answering tasks. Given the same constraints of a language for which there are not as many resources available and which is distant from English, ParsBERT [12] is created for Persian language.

Considering the Slavic group at which Macedonian language belongs, those languages are also considered as low-resource languages. There are several models for Polish language [10] among which is also multilingual SlavicBERT [4] for Russian, Bulgarian, Czech and Polish. SlavicBERT has shown that the languages with Latin script (Polish and Czech) performed better that the Cyrillic-based ones (Russian and Bulgarian), and there are multiple causes. The first is the dataset imbalance, the second is the incorrect sentence tokenization English sentence tokenizer has been used for Bulgarian, and third Russian and Bulgarian are much less related than Czech and Polish, thus the gain from having additional multilingual data is less. SlovakBERT [26] is the first Slovak only transformer model. The model was trained with 19.53GB of text data from various sources including: Wikipedia, Open Subtitles, OSCAR Corpus and self Web-crawled corpus. CroSloEngual BERT [32] is a trilingual model trained for Croatian, Slovenian and English language, and BERTić [20] is another model extending CroSlo-Engual BERT for Bosnian, Montenegrin and Serbian, beside the Croatian. As last from the Slavic group, there is also BERT model for Ukrainian language [19], however, to the best of our knowledge, the following languages remained unBERTed: Silesian, Kashubian, Macedonian, Serbian, Belarusian, and Rusyn.

This paper presents MACEDONIZER, the first BERT-based language model for Macedonian language.

3   MACEDONIZER: A Pre-trained ROBERTa Model in Macedonian

In this section, we are presenting the used pretraining dataset, objective and optimization part due the training of MACEDONIZER model.

3.1   Training Dataset

The dataset used for training the MACEDONIZER model is nearly 6.5 GB in total and is comprised of multiple subsets:

- News corpus representing texts until 2015 from the largest news aggregator in N. Macedonia - www.time.mk, with size of 4.95 GB;
- Macedonian corpus obtained from https://oscar-corpus.com with the size of 1.1 GB, and
- Macedonian Wikipedia dump obtained from https://dumps.wikimedia.org/mkwiki/ which is 373.3 MB in size.

3.2 Pre-training Objective

RoBERTa's base model architecture consists of 123 million parameters in the layers of the encoder, which means that is a fine-tuned BERT model, modified mainly in the hyperparameters. Additionally, RoBERTa shares the same architecture as BERT but employs a different pretraining strategy and a byte-level BPE as a tokenizer. We made the pre-training process on the masked language modeling task. This is actually the basic task of the transformer models, where 15% of the tokens are masked and predicted due the learning of the corpora. First of all, in order to train a transformer model in Macedonian or any other language which is not already covered, it is necessary to pre-train a RoBERTa Tokenizer for the specific language. In that purpose we pre-trained a RoBERTa tokenizer for Macedonian language. Our vocabulary consists 50 265 unique words and embedding size of 512. This pre-trained tokenizer is then used for preprocessing the input texts. Further, the model was trained on 1000 epochs, with 512 input vector size and 16 batch size. Other hyperparameters used are Adam Optimizer, 0.00001 learning rate, and 1 for gradient accumulation. The training of the model was done using Google Colab Pro with 1 GPU instance and 54GB RAM.

Figure 1 shows the loss curve during the pre-training of the model for each training step.

The training process lasted nearly two months, after which the evaluation of the model was done and referenced the performance with already existing multi-lingual transformer model. We then compared the results with the XLM-RoBERTa which contains an inner representation of 100 languages, among which is the Macedonian language. This model is used as a reference model for creating our MACEDONIZER. The main challenge was to surpass the success of XLM-RoBERTa at the downstream tasks.

4   Downstream Evaluation Tasks

MACEDONIZER is evaluated on three downstream tasks, represented in the following subsections: Named Entity Recognition (NER), Sentiment Analysis (SA) and Natural Language Inference (NLI). In addition, we give the baselines which will be used for comparison.

– **Sentiment Analysis (SA)** [23] is a field of text mining research that is continually developing. The algorithmic handling of text's views, feelings, and subjectivity is referred to as SA. The sentiment of the sentence depends on their type, such as: positive, negative or neutral sentiment. It can be a type of opinion, feeling or as we take the financial opportunity. We constructed the dataset for this downstream task, with collecting news from https:// time.mk/ and manually labeling it with the help of financial experts in two (Pos/Neg) classes. Our sentiment is not based on human emotions, otherwise it's positive or negative finance opportunity on the market based on the news, labeled with values 0 or 1 respectively. The data is split into three disjoint sets: train, validation, and test.

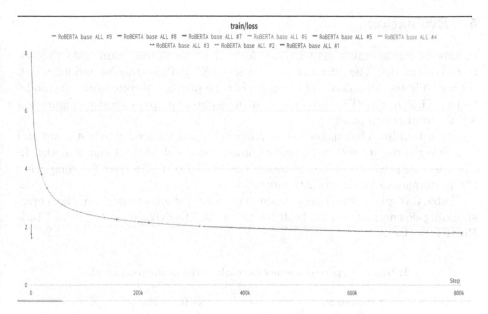

Fig. 1. Training loss curve

- **NER - Named Entity Recognition (NER)** [25] downstream task is based on highlighting entities in given sentences. NER is a task where a descriptive tag is declared for each word in a sentence. There are NER datasets and models that recognize different types of entities. The dataset[1] WikiANN (also known as PAN-X) is a multilingual named entity recognition dataset comprised of Wikipedia articles tagged in the IOB2 format with LOC (place), PER (person), and ORG (organisation) tags. This version conforms to Rahimi et al. (2019)'s balanced train, dev, and test splits, which cover 176 of the original WikiANN corpus's 282 languages [7]. The dataset is composed of 10k train, 1k for validation and 1k for testing.
- **Natural Language Inference (NLI)** [21]: the problem of deciding whether a natural-language hypothesis can be inferred from a given premise is a required (but not sufficient) condition for genuine natural language comprehension. Indeed, NLI may help with more urgent tasks like semantic search and question answering. For this task we translated part of the SNLI dataset[2] using Google Translator and then manually checked each sentence for validity of the translation. The SNLI data is a collection of 570k human-written English phrase pairs that have been manually tagged for balanced categorization with the labels entailment, contradiction, and neutral to aid natural language inference (NLI), also known as textual entailment recognition (RTE). We translated only 60k samples from the whole collection of the SNLI database, which we splitted in three parts, where we use 50k for training, 5k for validating and 5k for testing the model.

---

[1] https://huggingface.co/datasets/wikiann.
[2] https://huggingface.co/datasets/snli.

5 Evaluation

In this section are given explanations for each of the downstream tasks that we tested our model. The reference point was the XLM-Roberta-Base model, which is trained in the Macedonian Language. For the purpose of comparing our model to the XLM-RoBERTa model we used the same hyperparameter configuration for all downstream tasks.

After training a language-specific language model we needed to test it out and compere the results with an already known state-of-the-art language model. In this case, we used as a reference model the XLM-RoBERTa-Base for comparing the performances of all downstream tasks.

Table 1 displays the hyperparameters used for fine-tuning on the corresponding downstream tasks, both for the MACEDONIZER and for the XLM-RoBERTa-Base.

Table 1. Hyperparameters for each of the downstream tasks

Parameters	SA	NER	NLI
Epochs	10	5	5
Train-batch size	8	20	16
Validation-batch size	4	16	16
Learning rate	0.00001	0.00001	0.00001
Input sequence len.	256	128	250
Optimizer	Adam	Adam	Adam
Number of labels	2	7	3

The results from the evaluation are given in Table 2, where is shown that our model is performing better, according to the MCC score as the most appropriate metric for classification problems.

Table 2. Evaluation results on three different downstream tasks

Task	Model	Accuracy	Precision	Recall	F1	MCC
SA	XLM-RoBERTa-base	0.856	0.833	0.926	0.854	0.709
SA	**MACEDONIZER**	0.907	0.881	0.963	0.906	0.814
NER	XLM-RoBERTa-base	0.940	0.931	0.940	0.932	0.740
NER	**MACEDONIZER**	0.942	0.940	0.942	0.941	0.818
NLI	XLM-RoBERTa-base	0.652	0.653	0.652	0.652	0.480
NLI	**MACEDONIZER**	0.772	0.773	0.772	0.772	0.658

6 Discussion

After reviewing and analyzing the results, we can say that the Matthews correlation coefficient (MCC) yields a high score only if the forecast is correct in all four areas of the confusion matrix (true positives, false negatives, true negatives, and false positives) this confirms that our model is 10.5% better in the classification of sentiment analysis in financial news in Macedonian.

On the other hand for the NER downstream task our model has little improvement for the first metric. But, the MCC score shows that MACEDONIZER is 8.3% more precise in recognizing entities then XLM-RoBERTa-base model.

As for the last downstream task Natural Language Inference, the accuracy, precision, recall and F1-score are improved for 12% and the MCC score is 17.8% better with the use of MACEDONIZER. This means, that our model is in 17.8% of the cases better in recognizing the dependence between the hypothesis and the premise.

These percentages prove that models that are trained in a specific language often bring better results. But the dataset makes the risk of whether the model will prove successful. Through our research we determined that it is best if the corpus chosen is quite colorful and balanced. At the same time, it will be good if the vocabulary of the downstream tasks corresponds with the training one, which means that if it is a question of a model trained in a spoken language, it should not be tested only in a literary texts, which would lead to worse performance.

Lastly, we can proudly tell that MACEDONIZER is till now the best Macedonian Language Transformer in comparison for the improvement in every Downstream task that we evaluated.

7 Conclusion

This paper presents our efforts to pre-train the first contextual language model in the Macedonian language (MACEDONIZER) using the cutting-edge Transformer architecture. Our starting point is a randomly initialized RoBERTa model, which we pre-train using a textual corpus in the Macedonian language of approximately 6.5 GB. We use the self-supervised Masked Language Modeling (MLM) as a pre-training objective, where we mask 15% of the tokens. The number of epochs used in the pre-training phase is 1000. Next, the pre-trained version of the model is evaluated on three different downstream tasks: Sentiment Analysis (SA), Natural Language Inference (NLI) and Named Entity Recognition (NER). The evaluation results are compared to the cross-lingual version of the RoBERTa model - XML-RoBERTa, which supports more than one hundred languages, including Macedonian. The evaluation results show that MACEDONIZER achieves better scores than XML-RoBERTa in all downstream tasks, thus making our model the current state-of-the-art contextual language model in the Macedonian language. The pre-trained version of the MACEDONIZER is publicly available and distributed with the MIT license via HuggingFace[3].

---

[3] https://huggingface.co/macedonizer/mk-roberta-base.

Acknowledgement. The work in this paper was partially financed by the Faculty of Computer Science and Engineering, Ss. Cyril and Methodius University in Skopje.

References

1. Al-Garadi, M.A., et al.: Text classification models for the automatic detection of nonmedical prescription medication use from social media. BMC Med. Inform. Decis. Mak. **21**(1), 1–13 (2021)
2. Antoun, W., Baly, F., Hajj, H.: AraBERT: transformer-based model for Arabic language understanding. In: Proceedings of the 4th Workshop on Open-Source Arabic Corpora and Processing Tools, with a Shared Task on Offensive Language Detection, Marseille, France, pp. 9–15. European Language Resource Association (2020). https://aclanthology.org/2020.osact-1.2
3. Araci, D.: FinBERT: financial sentiment analysis with pre-trained language models. CoRR abs/1908.10063 (2019). http://arxiv.org/abs/1908.10063
4. Arkhipov, M., Trofimova, M., Kuratov, Y., Sorokin, A.: Tuning multilingual transformers for language-specific named entity recognition. In: Proceedings of the 7th Workshop on Balto-Slavic Natural Language Processing, pp. 89–93 (2019)
5. Brown, T.B., et al.: Language models are few-shot learners. CoRR abs/2005.14165 (2020). https://arxiv.org/abs/2005.14165
6. Carmo, D., Piau, M., Campiotti, I., Nogueira, R., de Alencar Lotufo, R.: PTT5: pretraining and validating the T5 model on Brazilian Portuguese data. CoRR abs/2008.09144 (2020). https://arxiv.org/abs/2008.09144
7. Chung, H.W., Garrette, D., Tan, K.C., Riesa, J.: Improving multilingual models with language-clustered vocabularies. In: Proceedings of the 2020 Conference on Empirical Methods in Natural Language Processing (EMNLP), pp. 4536–4546. Association for Computational Linguistics (2020). https://doi.org/10.18653/v1/2020.emnlp-main.367. https://aclanthology.org/2020.emnlp-main.367
8. Clark, K., Luong, M., Le, Q.V., Manning, C.D.: ELECTRA: pre-training text encoders as discriminators rather than generators. CoRR abs/2003.10555 (2020). https://arxiv.org/abs/2003.10555
9. Conneau, A., et al.: Unsupervised cross-lingual representation learning at scale. CoRR abs/1911.02116 (2019). http://arxiv.org/abs/1911.02116
10. Dadas, S., Perełkiewicz, M., Poświata, R.: Pre-training polish transformer-based language models at scale. In: Rutkowski, L., Scherer, R., Korytkowski, M., Pedrycz, W., Tadeusiewicz, R., Zurada, J.M. (eds.) ICAISC 2020. LNCS (LNAI), vol. 12416, pp. 301–314. Springer, Cham (2020). https://doi.org/10.1007/978-3-030-61534-5_27
11. Devlin, J., Chang, M., Lee, K., Toutanova, K.: BERT: pre-training of deep bidirectional transformers for language understanding. CoRR abs/1810.04805 (2018). http://arxiv.org/abs/1810.04805
12. Farahani, M., Gharachorloo, M., Farahani, M., Manthouri, M.: ParsBERT: transformer-based model for Persian language understanding. CoRR abs/2005.12515 (2020). https://arxiv.org/abs/2005.12515
13. He, P., Liu, X., Gao, J., Chen, W.: DeBERTa: decoding-enhanced BERT with disentangled attention. CoRR abs/2006.03654 (2020). https://arxiv.org/abs/2006.03654
14. Lample, G., Conneau, A.: Cross-lingual language model pretraining. CoRR abs/1901.07291 (2019). http://arxiv.org/abs/1901.07291

15. Lan, Z., Chen, M., Goodman, S., Gimpel, K., Sharma, P., Soricut, R.: ALBERT: a lite BERT for self-supervised learning of language representations. CoRR abs/1909.11942 (2019). http://arxiv.org/abs/1909.11942
16. Le, H., et al.: FlauBERT: unsupervised language model pre-training for French. CoRR abs/1912.05372 (2019). http://arxiv.org/abs/1912.05372
17. Lewis, M., et al.: BART: denoising sequence-to-sequence pre-training for natural language generation, translation, and comprehension. CoRR abs/1910.13461 (2019). http://arxiv.org/abs/1910.13461
18. Liu, Y., et al.: RoBERTa: a robustly optimized BERT pretraining approach (2019). https://doi.org/10.48550/ARXIV.1907.11692. https://arxiv.org/abs/1907.11692
19. Livinska, H.V., Makarevych, O.: Feasibility of improving BERT for linguistic prediction on Ukrainian corpus. In: COLINS (2020)
20. Ljubešić, N., Lauc, D.: Bertić-the transformer language model for Bosnian, Croatian, Montenegrin and Serbian. In: Proceedings of the 8th Workshop on Balto-Slavic Natural Language Processing, pp. 37–42 (2021)
21. MacCartney, B.: Natural Language Inference. Stanford University (2009)
22. Martin, L., et al.: CamemBERT: a tasty French language model. CoRR abs/1911.03894 (2019). http://arxiv.org/abs/1911.03894
23. Medhat, W., Hassan, A., Korashy, H.: Sentiment analysis algorithms and applications: a survey. Ain Shams Eng. J. **5**(4), 1093–1113 (2014)
24. Mishev, K., Gjorgjevikj, A., Vodenska, I., Chitkushev, L.T., Trajanov, D.: Evaluation of sentiment analysis in finance: from lexicons to transformers. IEEE Access **8**, 131662–131682 (2020)
25. Nadeau, D., Sekine, S.: A survey of named entity recognition and classification. Lingvisticae Investigationes **30**(1), 3–26 (2007)
26. Pikuliak, M., et al.: SlovakBERT: Slovak masked language model. CoRR abs/2109.15254 (2021). https://arxiv.org/abs/2109.15254
27. Radford, A., Wu, J., Child, R., Luan, D., Amodei, D., Sutskever, I., et al.: Language models are unsupervised multitask learners. OpenAI Blog **1**(8), 9 (2019)
28. Raffel, C., et al.: Exploring the limits of transfer learning with a unified text-to-text transformer. CoRR abs/1910.10683 (2019). http://arxiv.org/abs/1910.10683
29. Rasmy, L., Xiang, Y., Xie, Z., Tao, C., Zhi, D.: Med-BERT: pretrained contextualized embeddings on large-scale structured electronic health records for disease prediction. NPJ Digit. Med. **4**(1), 1–13 (2021)
30. Sanh, V., Debut, L., Chaumond, J., Wolf, T.: DistilBERT, a distilled version of BERT: smaller, faster, cheaper and lighter. CoRR abs/1910.01108 (2019). http://arxiv.org/abs/1910.01108
31. Souza, F., Nogueira, R., Lotufo, R.: BERTimbau: pretrained BERT models for Brazilian Portuguese. In: Cerri, R., Prati, R.C. (eds.) BRACIS 2020. LNCS (LNAI), vol. 12319, pp. 403–417. Springer, Cham (2020). https://doi.org/10.1007/978-3-030-61377-8_28
32. Ulčar, M., Robnik-Šikonja, M.: Finest BERT and crosloengual BERT: less is more in multilingual models. CoRR abs/2006.07890 (2020). https://arxiv.org/abs/2006.07890
33. Vaswani, A., et al.: Attention is all you need. In: Advances in Neural Information Processing Systems, pp. 5998–6008 (2017)
34. de Vries, W., Nissim, M.: As good as new. How to successfully recycle English GPT-2 to make models for other languages. CoRR abs/2012.05628 (2020). https://arxiv.org/abs/2012.05628

35. Yang, Z., Dai, Z., Yang, Y., Carbonell, J., Salakhutdinov, R.R., Le, Q.V.: XLNet: generalized autoregressive pretraining for language understanding. In: Advances in Neural Information Processing Systems, vol. 32 (2019)
36. Zhu, Y., et al.: Aligning books and movies: towards story-like visual explanations by watching movies and reading books. In: Proceedings of the IEEE International Conference on Computer Vision, pp. 19–27 (2015)

Deep Learning-Based Sentiment Classification of Social Network Texts in Amharic Language

Senait Gebremichael Tesfagergish[1]([⊠]), Robertas Damaševičius[1], and Jurgita Kapočiūtė-Dzikienė[2]

[1] Kaunas University of Technology, 51368 Kaunas, Lithuania
sengeb@ktu.lt
[2] Vytautas Magnus University, 44404 Kaunas, Lithuania

Abstract. Sentiment analysis is among the main targets of natural language processing (NLP) that assigns a positive or negative value to the opinion expressed in natural language text within different contexts such as social media, forum, news, blogs, and many others. Sentiments of an under-researched language such as Amharic have received little attention in NLP applications due to the scares of enough resources to develop such methods. In this paper we combine the deep learning (CNN, LSTM, FFNN, and BiLSTM) and classical models (cosine similarity) with word embedding techniques for sentence-level sentiment classification of social media messages in Amharic language that has never been tested before. We use the Amharic Twitter dataset that consists of around 3000 text snippets. Data augmentation is applied to increase the dataset for training those models. We achieved the 82.2% accuracy using the sentence transformer and cosine similarity on the Amharic corpus.

Keywords: Deep learning · Sentiment analysis · Sentence transformer · Amharic language

1 Introduction

Sentiment analysis is a process of analyzing text to detect overall sentiment toward topic-positive, negative, mixed, and neutral. The main purpose of analyzing text and identifying their sentiments are relies on the technological era we live today. Everything is shifting to online and online comments and reviews from the end users affect the decision taken by stakeholders in different domains [1]. In stock markets: the movements of stock market are influenced by messages on social networks [2]. News with a generally favorable tone has been linked to a significant price increase. Negative news, on the other hand, is thought to be connected to a price drop with longer-term consequences. In entertainment industry, customer reviews and comments are used for decision making for other potential buyers of the products [3]. Similarly, producers use it for improving the quality of their service and outline a plan for their coming products or services. In politics, it helps authorities to make decisions based on the overall sentiment from population surveys [4]. In healthcare, analysis of online comments and text messages can be used for medical

K. Zdravkova and L. Basnarkov (Eds.): ICT Innovations 2022, CCIS 1740, pp. 63–75, 2022.
https://doi.org/10.1007/978-3-031-22792-9_6

diagnostics of diseases [5]. A dark side of social networks is that they can be used to spread hate speech [6] and disinformation (aka "fake news") [7], aiming to influence the events in real world.

Due to ambiguities in each language and our human understanding there is no single solution that could work for all languages: each language is different and difficult in its own way, therefore requires adaptation [8]. The analysis of morphological features of a language is needed for real-world natural language processing (NLP) tasks [9]. Under-researched languages like Amharic [10] could not get the benefit of the application and tools that already developed for the resource rich languages. It is due to its morphological complexity and unavailability of enough data for solving the sentiment analysis [11] task. Innovative artificial intelligence (AI) methods are helping the under-resourced languages to pass the hardships of collecting and preprocessing of large datasets, instead they provide a deep insight of the available data features to make the classification more efficient [12].

The aim of this work is to get the benefit of the state-of-the art deep learning-based solutions for the sentiment analysis task for Amharic language in a sentence level. The main novelty and contribution of this experiment is outlined as follows:

- The state-of-the art sentence transformer embedding is applied to Amharic language sentiment classification for the first time.
- Different deep learning (FFNN, CNN, LSTM, BILSTM) and classical (cosine similarity and KNN) machine learning classifiers are tested and compared with Word2Vec and sentence embeddings.
- Classical machine learning methods, Cosine Similarity and K-nearest neighborhood are used as hybrid model and tested for the first time in Amharic Sentiment classification task.

This paper is structured as follows. Related works are described in Sect. 2. The dataset used for this experiment are presented in Sect. 3. Analysis of vectorization, classification models and optimization techniques are discussed in Sect. 4. Section 5 explains the experiment and its results. Section 6 concludes with the discussion and conclusion about overall objectives and achievements of this study and future works.

2   Related Works

Amharic language and many other low-resourced Semitic languages are neglected from the NLP research due to the scarce of electronic data and basic tools needed to perform such experiments. Amharic language is the 2nd most commonly spoken Semitic language after Arabic. It is an official language of Ethiopia and it has around 22 million native speakers. Amharic NLP research is fairly conducted more than the other Semitic languages, and it is supported by Google translation.

In this section the focus will be on the related and recent achievements of sentiment analysis task in Amharic and Arabic. Although Amharic NLP research needs more research and investigation, due to its morphological complexity several studies were conducted. Recent study [13] focused in building an annotation tool and investigated several classical, and deep learning classification methods. FLAIR deep learning text classifier based on the graphical embedding achieves the highest accuracy which is 54.53%. Authors concluded that deep learning methods are more suitable to tackle the sentiment analysis task more than the classical machine learning methods.

Another deep learning approach for Amharic [14] used data extracted from the official Facebook page of Fana broadcasting news media in Ethiopia. The experiment used the TF-IDF count vectorization method and deep learning architecture. The best achievement registered was 90.1%. A study [15] used the Naïve Bayes algorithm with unigrams, bigrams and hybrid features. The research was conducted on 600 posts labelled to two classes: positive and negative. The authors managed to get their highest result at 44% using the bigram feature. Bert [16] was also tested for classification of 2 class sentiment analysis using 6652 samples collected from Facebook comments. Authors finetune the pretrained BERT model and achieved 95% accuracy.

Arabic language shared many similar characteristics with Amharic in terms of morphology. Morphology of vowel-consonant templates for forming verbs and nouns and, they both are inflected for gender, case, and number. They also share many words and vocabularies. Sentiment analysis of Arabic tweets [17] classified into four classes. They used pre-trained word embeddings trained using the Word2Vec. The highest accuracy of 64.75%, was achieved using LSTM, outperforming the CNN model by 0.4%.

The CNN-LSTM method presented in [18] classify sentiment into two classes using the dataset consists of 63,000 reviews dataset. Applying the method in the top of the fast test embedding achieves the best result of 90.75%.

Summarizing, previous works used classical supervised machine learning approaches like SVM, multinomial Naïve Bayes, Maximum Entropy algorithms using Bag of Words, and Naïve Bayes decision tree for document level Amharic sentiment classification. The latest research done for sentiment analysis is performed using deep leaning methods due to their ability to outpace traditional methods. However, sentiment analysis task is a sequence task and other methods like RNN, CNN are more efficient for solving problem involving sequence dependencies. In recent years, deep learning techniques such as CNNs and recurrent neural networks (RNNs) have captured the attention of researchers in the sentiment analysis task [19]. They take the dependencies of the words in given sentence into account, which is important since word positioning can make a notable difference in the meaning of the sentence.

3 Dataset

The dataset used for the sentiment analysis task is the Ethiopic Twitter Dataset for Amharic (ETD-AM) introduced by Yimam et al. [20]. The data is acquired from Twitter and annotated using the Amharic Sentiment Annotator Bot (ASAB) [13]. ETD-AM stores only tweet ids and their sentiments. For retrieving tweets, the tweepy python library is used for accessing the Twitter API.

The original dataset consists around 9.4K tweets mapped into four classes (Positive, Negative, Neutral and Mixed). We omitted the Neutral and Mixed classes from this experiment. In addition, some tweets were not returned by the API calls. As a result, 1736 negative and 1516 positive tweets were used for further experiments.

After downloading the tweets, it needs to go through several pre-processing before vectorization. Social network texts are known to contain numerous spelling mistakes, slang phrases, and multilingual words, particularly in blogs, Twitter, and online conversations. As a result, preprocessing is utilized to eliminate unnecessary content and convert it into a readable format to extract the emotion as follows. Cleaning: The original dataset is a collection of tweets, so as it consists non-Geez script. Therefore Emojis, web links, Latin letters and English words are removed. Tokenization: the text is divided into tokens using the 'Tokenizer' from the Python Keras library.

Dataset size has a big impact in the quality of the trained model. A larger dataset with good quality data will allow to train a model with better accuracy. In our case the dataset is small, so we needed to augment it with more data. The English dataset from Twitter (Sentiment140) [21] was translated to Amharic and added to original dataset. The English dataset is a very popular sentiment annotated dataset of 1.6 million tweets. The added translated data is balanced where positive class has 15,000 instances and negative class has also 15,000. For translation we use the Google translator.

4 Methods

4.1 Vectorization

Machine leaning classifiers can learn through the calculated difference, relationship and probability from the input value or data. Inputs for NLP tasks are usually natural language texts or scripts which is not possible to feed the inputs as it is into the machine learning classifiers. Therefore, vectorization is used to bridge the gap between the input and the ML classifiers for NLP tasks. Word embeddings are used to map each word as a low-dimensional vector. In this experiment the following types of wording embedding are tested.

Word2Vec is a neural based word embedding blends the skip gram and CBOW approaches to forecast the context of the input word in the real values vector space of given dimension. The quality of the embeddings is affected by the amount of data used to learn the network. Amharic does not have any publicly available pretrained Word2Vec embedding. Word2vec is trained using the Ethiopic Twitter Dataset for Amharic (ETD-AM) with 300 dimension and window size equal to 5.

Sentence Transformers are state-of-the-art technique that enables to derive semantically meaningful sentence embeddings [22]. They use the Siamese network that can generate fixed-sized vectors for an input sentence. Using a similarity metric such as Euclidean/Manhattan distance or cosine similarity, semantically similar sentences can be found. This framework can be used for sentence embedding in more than 100 languages. The Amharic language has the pretrained model of language-agnostic BERT sentence embedding model (LaBSE) [23], and we used it for our experiment.

4.2 Classification Models

In this study, sentiment classification for Amharic language is a binary classification task to assign either positive or negative class to a given sentence. Several classical and deep learning methods are trained and tested to tackle this task as follows.

Feed Forward Neural Network (FFNN) is a simple deep learning classifier used when nonlinear mapping is done between inputs and outputs to predict future state. FFNN is used on the top of the embeddings from the sentence transformers. The model (Fig. 1) is trained to learn the relationship in between sentences from the embeddings. When testing, it predicts the class of the most similar sentence in the training set.

```
Classifier(
  (dp): Dropout(p=0.1, inplace=False)
  (ff): Linear(in_features=768, out_features=2, bias=True)
)
```

Fig. 1. Architecture of FFNN model.

Convolutional Neutral Network (CNN) was originally developed to image processing and classification but recently it successfully adapted to text classification. First, it performs feature extraction from the given embeddings (sentences/word) and classification using the 1D convolutional layer. CNN only considers a pattern of sequential words (called n-grams), in which it can be unsuitable for sentiment analysis task as a word out of the n-gram window can change the meaning of the sentence. However, we still used this model (Fig. 2) as a baseline for comparison.

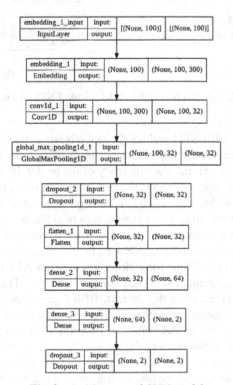

Fig. 2. Architecture of CNN model

RNNs such as Long Short-Term Memory (LSTM) and Bi-directional LSTM (BiL-STM) work well with sequential data as it takes the whole sequence of words into consideration while training the model. The models (Fig. 3) use the output of the previous state as input for a current state. Its limitation is that this network is suffering from the vanishing gradient problem. For this problem the LSTM model is introduced as a solution to identify which information to input, forget and output gates. BiLSTM is a modification of the LSTM model, which learns from two directions of the sentence (from beginning to end and from end to beginning).

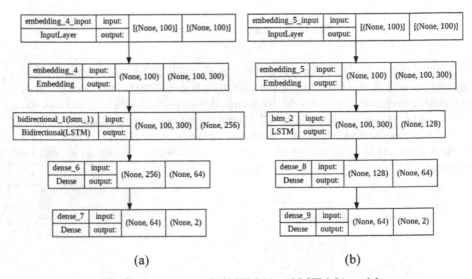

(a)   (b)

Fig. 3. Architecture of BiLSTM (a) and LSTM (b) model

The hybrid (CNN & BiLSTM and LSTM & CNN) models blend different architectures to solve a classification task gives the opportunity to use those unique characteristics of those architecture to maximize the efficiency of the model. In this study, we blended two parts for the hybrid model (Fig. 4). The first part is the CNN model, and the second part is BiLSTM and LSTM. The idea behind this pairing is that the first component of the model extracts features while the second component learns from the input text [24].

Bidirectional Encoder Representations from Transformer (BERT) is a transformer-based technique for NLP pre-training developed by Google. Its generalization capability is such that it can be easily adopted for various down-stream NLP tasks such as question answering, relation extraction, or sentiment analysis [25, 26]. Transformers are used to learn the relationship of words in the context. BERT generates language model using the encoder. The bidirectional encoder reads the sequence in both directions (right-to-left and left-to-right), so the model is trained from the right and left side of the target word. Because the core architecture is trained on a huge text corpus, the parameters of the architecture's most internal levels remain fixed. The outermost layers, on the other hand, adapt to the job and are where fine-tuning takes place. Sentiment analysis is done by adding a final classification layer to process the transformer output for the [CLS] token. Currently, the Amharic pre-trained BERT model is not available yet. Therefore, the English model was adapted.

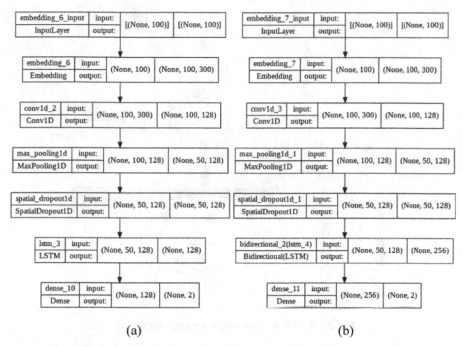

(a)   (b)

Fig. 4. Architecture of hybrid CNN-BiLSTM (a) and CNN-LSTM (b) models

Cosine Similarity is measuring the cosine value of the angle between two sentences (vectors), it determines if the two sentences are pointing into the same direction, so they are similar. It is well known procedure to identify document similarity. The calculated value is given in the range of 0 to 1, that means two same sentences gives the cosine similarity equal to 1. In this experiment, once a testing sentence is given the model compute the cosine distance between the testing sentence and all the sentence in the training set and output the most similar sentence's class with the highest distance measure. Cosine similarity is applied on the top of sentence transformer.

K-nearest neighborhood (KNN) is a classification algorithm that assigns a class label to an unknown data sample based on the class labels of the closest known data samples. In this study, KNN is blended with the cosine similarity to group the K-nearest sentences in the training set to the test sentence. After grouping the K-nearest neighborhood sentences the class of the group is assigned by the highest occurrence of the class in the group, and the testing sentence will get the same class as the group.

Our methodology is summarized in Fig. 5.

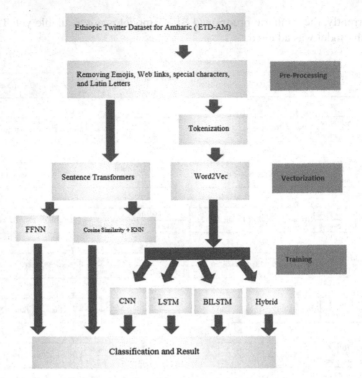

Fig. 5. Workflow of methodology and experiment

5 Experiment and Results

Sentiment analysis is a task to classify the sentiments written in natural language into positive negative, or neutral classes. Tweets in the ETD-AM dataset consist of one or two sentences, which are bounded to the largest length of messages set by Twitter. Word embeddings are used as a feature for the classification for this experiment.

The dataset is described in Sect. 3 are vectorized by different embedding techniques explained in Sect. 4.1 and classified using the methods in explained in Sect. 4.2. For the implementation of the methods of Tensorflow, Keras, and PyTorch libraries with the python language implementation is used.

The results of different classifiers using original and augmented datasets are presented in Table 1. Note that adding more data using sentences translated from English dataset improves the results that used the Word2Vec and deep learning methods (CNN, BiLSTM, CNN-LSTM, CNN-BiLSTM), while it downgrades (by 5%) the performance of the best model (with 82% accuracy) that uses the Sentence Transformers (FFNN and Cosine Similarity + KNN). In this case the possible reason for this result can be the domain of the texts. The domain of the original dataset was Ethiopian politics, whereas the domain of the translated English dataset was randomly extracted tweets from Twitter of different domains.

Table 1. Accuracy of original data and accuracy with added translated data

Model	Accuracy (original dataset)	Accuracy (augmented dataset)
CNN + Word2Vec	0.46	0.64
LSTM + Word2Vec	0.54	0.49
BILSTM + Word2Vec	0.62	0.68
CNN & BILSTM + Word2Vec	0.41	0.69
CNN & LSTM + Word2Vec	0.39	0.70
Sentence Transformer + Cosine Similarity + KNN	**0.82**	0.77
Sentence Transformer + FFNN	0.80	0.76
BERT	–	0.54

The KNN classifier is used in the best model has a hyperparameter of the number of nearest neighbors. To explore the possible values of this hyperparameter, we performed the ablation study and present its results in Table 2. The best accuracy was achieved with 157 nearest neighbors.

Table 2. Accuracy of Cosine Similarity with the K-nearest neighborhoods

Hyperparameter value (number of nearest neighbors (NN))	Accuracy of Sentence Transformer + Cosine Similarity + KNN model
1-NN	0.72
3-NN	0.78
31-NN	0.80
59-NN	0.81
157-NN	**0.82**

Finally, the Precision, Recall, F1-Score and Accuracy of all tested classification models are summarized in Table 3. The best performance was achieved by the hybrid Sentence Transformer (ST) + Cosine Similarity + KNN model (an accuracy of 82.1%). We also present the confusion matrix of the best model in Fig. 6.

Table 3. Precision, Recall, F1-score and Accuracy of all tested classification models

Model	Precision	Recall	F1-score	Accuracy
CNN + Word2Vec	0.65	0.57	0.60	0.64
LSTM + Word2Vec	0.27	0.50	0.35	0.54
BILSTM + Word2Vec	0.66	0.60	0.62	0.68
CNN & BILSTM + Word2Vec	0.72	0.62	0.67	0.69
CNN & LSTM + Word2Vec	0.69	0.73	0.71	0.70
ST + Cosine Similarity + KNN	**0.822**	**0.821**	**0.821**	**0.821**
ST + FFNN	0.806	0.799	0.801	0.804
BERT	–	–	0.462	0.539

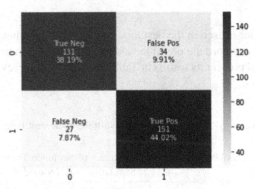

Fig. 6. Confusion matrix of best model using Sentence Transformer + Cosine Similarity + KNN

6 Discussion

In this study, we compared word vectorization and classification methods for sentiment analysis in Amharic. The vectorization methods are both at the word and sentence level. The classification methods also include the classical (KNN and Cosine Similarity) and deep learning (FFNN, LSTM, BiLSTM, CNN-LSTM, CNN-BiLSTM, BERT) machine learning methods. After testing different classification and vectorization models, the best result was achieved using the hybrid model of Cosine similarity and KNN with the Sentence transformer vectorization. Sentence transformers are the state-of-the art methods that give a good estimation of Amharic sentence vectors. Therefore, the classification methods on the top of the sentence transformers are not expected to be more complex as the features of the sentence is already learned well in the vectorization step. In our experiment we tested Sentence transformers with FFNN and Cosine Similarity + KNN, and the first and second ranked best results of the overall experiment were achieved by those methods that used Sentence Transformers.

The proposed method of Sentence transformer vectorization and Cosine similarity classifier has some limitations due to several factors: a small dataset for both classification and vectorization; Amharic language has morphological richness; absence of contextual information; ambiguity and dialects of Amharic; lack of pre-processing tools (lemmatization, part-of-speech tagging, normalization, stemming, etc.); low number of words in a sentence (short sentences was the highest misclassified texts).

7 Conclusion

Sentiment analysis is a popular NLP task consisting of grading opinions as positive, negative, neutral, or mixed. Its importance lies in that it automatically extracts the sentiment from messages originating from a large group of audience. Previous work used the same Amharic dataset tested the deep learning approaches, and it achieved the highest accuracy of 54.53%, when word2vec embedding is trained with 15 million sentences [13]. However, in this paper the best result achieved of hybrid model CNN & LSTM using 5.3K sentences for Word2Vec vectorization outperformed the previous research and achieved 70% accuracy. State-of-the-art sentence transformers are also tested for the first time on the top of feedforward neural network and Cosine Similarity. The highest accuracy is achieved by the cosine similarity with the K-nearest neighborhood voting technique, which is 82%. Data augmentation produced mixed results: it shows improvement in the deep learning models accuracy, while it reduces the accuracy in the experiment using the sentence transformers.

The domain in which the data set (sentiments) has a large influence on accuracy. In this experiment, the gold standard ETD Amharic dataset has Ethiopian political content, which makes it so hard to find similar data in English and translate. Data augmentation may be more efficient in increasing the accuracy when it has the same domain as the original dataset. The result of this experiment can benefit, other Semitic languages based in East Africa that shares similar morphological structure with Amharic (e.g., Tigrinya, Tigre).

Future work will focus on multiclass sentiment classification and include more information appears in the written opinion such as emoticons.

References

1. Ji, Z., Pi, H., Wei, W., Xiong, B., Wozniak, M., Damasevicius, R.: Recommendation based on review texts and social communities: a hybrid model. IEEE Access 7, 40416–40427 (2019). https://doi.org/10.1109/ACCESS.2019.2897586
2. Behera, R.K., Das, S., Rath, S.K., Misra, S., Damasevicius, R.: Comparative study of real time machine learning models for stock prediction through streaming data. J. Universal Comput. Sci. 26(9), 1128–1147 (2020)
3. Vaiciukynaite, E., Zailskaite-Jakste, L., Damasevicius, R., Gatautis, R.: Does hedonic content of brand posts affect consumer sociability behaviour on Facebook? In: Proceedings of the 5th European Conference on Social Media, ECSM 2018, pp. 325–331 (2018)
4. Okewu, E., Misra, S., Okewu, J., Damaševičius, R., Maskeliūnas, R.: An intelligent advisory system to support managerial decisions for a social safety net. Adm. Sci. 9(3), 55 (2019). https://doi.org/10.3390/admsci9030055

5. Omoregbe, N.A.I., Ndaman, I.O., Misra, S., Abayomi-Alli, O.O., Damaševičius, R.: Text messaging-based medical diagnosis using natural language processing and fuzzy logic. J. Healthc. Eng. **2020**, 1–14 (2020). https://doi.org/10.1155/2020/8839524
6. Aldjanabi, W., Dahou, A., Al-Qaness, M.A.A., Elaziz, M.A., Helmi, A.M., Damaševičius, R.: Arabic offensive and hate speech detection using a cross-corpora multi-task learning model. Informatics **8**(4), 69 (2021). https://doi.org/10.3390/informatics8040069
7. Tesfagergish, S.G., Damaševičius, R., Kapočiūtė-Dzikienė, J.: Deep fake recognition in tweets using text augmentation, word embeddings and deep learning. In: Gervasi, O., et al. (eds.) ICCSA 2021. LNCS, vol. 12954, pp. 523–538. Springer, Cham (2021). https://doi.org/10.1007/978-3-030-86979-3_37
8. Venčkauskas, A., Damaševičius, R., Marcinkevičius, R., Karpavičius, A.: Problems of authorship identification of the national language electronic discourse. In: Dregvaite, G., Damasevicius, R. (eds.) ICIST 2015. CCIS, vol. 538, pp. 415–432. Springer, Cham (2015). https://doi.org/10.1007/978-3-319-24770-0_36
9. Choi, M., Shin, J., Kim, H.: Robust feature extraction method for automatic sentiment classification of erroneous online customer reviews. Information (Japan) **16**(10), 7637–7646 (2013)
10. Gereme, F., Zhu, W., Ayall, T., Alemu, D.: Combating fake news in "low-resource" languages: amharic fake news detection accompanied by resource crafting. Information **12**, 20 (2021). https://doi.org/10.3390/info12010020
11. Nandwani, P., Verma, R.: A review on sentiment analysis and emotion detection from text. Soc. Netw. Anal. Min. **11**(1), 1–19 (2021). https://doi.org/10.1007/s13278-021-00776-6
12. Kapočiūtė-Dzikienė, J., Damaševičius, R., Woźniak, M.: Sentiment analysis of Lithuanian texts using traditional and deep learning approaches. Computers **8**(1), 4 (2019)
13. Yimam, S.M., Alemayehu, H.M., Ayele, A., Biemann, C.: Exploring amharic sentiment analysis for social media texts: building annotation tools and classification models. In: Proceeding of the 28th International Conference on Computational Linguistics (2020)
14. Getachew, Y., Alemu, A.: Deep learning approach for amharic sentiment analysis. University Of Gondar (2018)
15. Wondwossen, P., Wondwossen, M.: A machine learning approach to multi-scale sentiment analysis of amharic online posts. HiLCoE J. Comput. Sci. Technol. **2**(2), 8 (2014)
16. Neshir, G., Atnafu, S., Rauber, A.: BERT fine-tuning for amharic sentiment classification. In: Workshop RESOURCEFUL Co-Located with the Eighth Swedish Language Technology Conference (SLTC), Gothenburg, Sweden, 25 November 2020 (2020)
17. Heikal, M., Torki, M., El-Makky, N.: Sentiment analysis of Arabic tweets using deep learning. Proc. Comput. Sci. **142**, 114–122 (2018)
18. Ombabi, A.H., Ouarda, W., Alimi, A.M.: Deep learning CNN–LSTM framework for Arabic sentiment analysis using textual information shared in social networks. Soc. Netw. Anal. Min. **10**(1), 1–13 (2020). https://doi.org/10.1007/s13278-020-00668-1
19. Tang, D., Qin, B., Liu, T.: Deep learning for sentiment analysis: successful approaches and future challenges. Wiley Interdiscipl. Rev. Data Min. Knowl. Discov. **5**(6), 292–303 (2015)
20. Yimam, S.M., Ayele, A.A., Biemann, C.: Analysis of the ethiopic Twitter dataset for abusive speech in amharic. In: International Conference on Language Technologies for All: Enabling Linguistic Diversity And Multilingualism Worldwide, Paris, France, pp. 1–5 (2019)
21. Kaggle. Sentiment140 Dataset with 1.6 Million Tweets. https://www.kaggle.com/kazanova/sentiment140. Accessed 8 Jan 2022
22. Reimers, N., Gurevych, I.: Sentence-BERT: sentence embeddings using Siamese BERT-networks. In: 2019 Conference on Empirical Methods in Natural Language Processing and the 9th International Joint Conference on Natural Language Processing (EMNLP-IJCNLP) (2019)

23. Feng, F., Yang, Y., Cer, D., Arivazhagan, N., Wang, W.: Language-agnostic BERT sentence embedding. In: Proceedings of the 60th Annual Meeting of the Association for Computational Linguistics (Volume 1: Long Papers) (2022). https://doi.org/10.18653/v1/2022.acl-long.62
24. Pota, M., Ventura, M., Catelli, R., Esposito, M.: An effective BERT-based pipeline for twitter sentiment analysis: a case study in Italian. Sensors **21**(1), 1–21 (2021)
25. Go, A., Bhayani, R., Huang, L.: Twitter sentiment classification using distant supervision. CS224N Project Report, Stanford, 1(2009), p. 12 (2009)
26. Tesfagergish, S.G., Kapočiūtė-Dzikienė, J., Damaševičius, R.: Zero-shot emotion detection for semi-supervised sentiment analysis using sentence transformers and ensemble learning. Appl. Sci. **12**, 8662 (2022). https://doi.org/10.3390/app12178662

Using Centrality Measures to Extract Knowledge from Cryptocurrencies' Interdependencies Networks

Hristijan Peshov[1(✉)], Ana Todorovska[1], Jovana Marojevikj[1], Eva Spirovska[1], Ivan Rusevski[1], Gorast Angelovski[1], Irena Vodenska[1,3], Ljubomir Chitkushev[2], and Dimitar Trajanov[1,2]

[1] Faculty of Computer Science and Engineering, Ss. Cyril and Methodius University in Skopje, Skopje, Republic of North Macedonia
`hristijan.peshov@students.finki.ukim.mk`
[2] Computer Science Department, Metropolitan College, Boston University, Boston, MA, USA
[3] Administrative Sciences Department, Financial Management, Metropolitan College, Boston University, Boston, MA, USA

Abstract. Is the rising price of Bitcoin affected by Ethereum's fall? Are cryptocurrencies interconnected and are shifts in prices a consequence of said influence, or maybe social media plays a more significant role? To answer these questions, we create 7 networks using different approaches, each of them representing the relationship between 18 most popular cryptocurrencies in a distinct way. Additionally, by calculating centrality measures on the networks, we discover the currency that will be the first to spread their influence onto others. Moreover, these measures detects a currency with a high influence over the entire network, as well as the one that have the most "important" neighbors. Our results show that cryptocurrencies are indeed interrelated, especially the more popular ones, which also happens to be the most affected by the social media platforms. Ethereum is one of the fastest to affect the others when change in price occur, while both Ethereum and Bitcoin have extensive reach in the networks.

Keywords: Cryptocurrency · NLP · Sentiment · Networks · Centrality measures · Machine learning · Explainable AI

1 Introduction

As human beings, we have always been genetically driven to connect with other people. We connect in order to socialize, we work, live in interlinked communities. A human's life is filled with connections. Our brain, as well, is composed of links and relations that bind neurons. We gain deeper knowledge and remember things more easily when we make associations. Thus we seek links in everything. Consequently, as the cryptocurrencies' importance becomes increasingly

K. Zdravkova and L. Basnarkov (Eds.): ICT Innovations 2022, CCIS 1740, pp. 76–90, 2022.
https://doi.org/10.1007/978-3-031-22792-9_7

significant, researchers and investors are more interested in associations between them, as well as what causes their prices to shift. The new concept of virtual currencies called cryptocurrencies, poses a new connectivity challenge regarding their relationship with different economic assets. A cryptocurrency is a digital currency that uses cryptography to secure transactions. These currencies rely on blockchain technology that provides the security of transactions. It is noteworthy to point out that unlike the fiat money such as U.S. Dollar or the Euro, there is no central authority that maintains the value of the cryptocurrencies. They are decentralized assets.

Bitcoin is the first one, created by Satoshi Nakamoto in 2009 [1]. Today, Bitcoin is the leader in the world of cryptocurrencies in terms of popularity and value. It possesses the highest price for a single token, valued at over 45.000 USD at the start of 2022. In addition to this, the importance of cryptocurrencies is supported by the fact that there are over 20.000 cryptocurrencies as of July 2022[1], which clearly indicate the vast expansion of this field.

An important source of influence on the cryptocurrency price are the social media platforms such as Reddit and Twitter where a great amount of data circulates every minute. The impact is noticeable, especially when people are making financial decisions based on the content they are consuming on these platforms. For instance, a survey conducted among 1302 U.S. investors, claims that more than 1 in 5 have used Reddit to make an investment decision[2]. Another popular examples are tweets from Elon Musk, the Tesla CEO. The tweet, "One word: Doge" sent the shares of the cryptocurrency Dogecoin up nearly 20%, bringing the currency on the list of trending Twitter topics[3]. Musk drove 14% surge of the Bitcoin, when he added #Bitcoin hashtag to his Twitter bio[4]. Such a big shift in the cryptocurrencies' prices affected by the social media show us that the social platforms must be taken into consideration when analyzing the performance of cryptocurrencies. To explore these connections we use different approaches where the Reddit posts or Google News information are considered when examining the cryptocurrency behavior.

One drawback in researching cryptocurrency behavior is that they are exceptionally volatile. Accordingly, prices are constantly varying, thus making it difficult to provide a forecast, which gives investors a hard time. Instead of trying to discover a pattern in the crypto charts and predicting the price of a single currency, we focus on finding relations between the cryptocurrencies. We use multiple approaches with the aim of expressing one's influence on the others. To achieve this, we construct 7 networks of cryptocurrencies representing different relationships between them [2]. In [2], we collected the data and created

---

[1] https://explodingtopics.com/blog/number-of-cryptocurrencies.

[2] https://www.investing.com/blog/beyond-gamestop-survey-reveals-reddits-largerthanexpected-influence-on-investing-298.

[3] https://edition.cnn.com/2020/12/20/investing/elon-musk-bitcoin-dogecoin/index.html.

[4] https://cointelegraph.com/news/elon-musk-adds-bitcoin-to-twitter-bio-with-43-7m-followers.

the networks that we use in this paper. However, in this paper, we extend that research by applying centrality measures on the networks. We use these measures to extract additional knowledge from the networks with the aim to discover the assets that are most beneficial to observe.

This paper is structured as follows. Section 2 presents the data gathered and used in this research. Section 3 elaborates the methodology that we use to obtain the networks, while centrality measures and the results are shown in Sect. 4. The last section, Sect. 5 concludes our findings.

2 Data

The more available data we have, the better the opportunities to research are. In order to make constructive decisions, we depend on data. The advances in Machine Learning (ML) and Artificial Intelligence (AI) boost the extraction of useful information from publicly available sources, especially social platforms. This information improves the prediction of cryptocurrencies' behavior. The gained knowledge used to obtain the networks and calculate centrality measures, contains cryptocurrency prices, Reddit posts' titles and Google News titles [2]. We use each of these sources both individually and combined as an input to the forecasting model.

2.1 Cryptocurrency Prices

We present 18 most dominant cryptocurrencies according to their popularity and appearance in Google Trends[5]. Our list consist of Ethereum, Bitcoin, Nano, EOS, Ripple, Stellar, Litecoin, NEM, Celsius, VeChain, Dash, Dogecoin, Chainlink, Maker, Monero, Cardano, Neo, and Iota. We obtain the cryptocurrencies data from Goldprice[6] and Yahoo finance[7].

The final dataset for each cryptocurrency contains date, high and low price, open price, volume, close price and adjusted closing price, for a time span of 2 years, from March 2019 to March 2021.

2.2 Reddit Posts' Titles

To obtain Reddit titles for the selected cryptocurrrencies, we apply a crawler[8] that searches for titles that contain a specific cryptocurrency name. Only those titles are analyzed in the further research. To extract this beneficial information, we examine the Reddit group "Cryptocurrency news & Discussion"[9], plus another eighteen groups, each of which is exclusive to a single cryptocurrency. We collect the titles together with an id, a timestamp, number of comments, score and link to the post. All collected data is written in English, for a time span of 2 years, from March 2019 to March 2021.

---

[5] https://trends.google.com/trends/?geo=US.
[6] https://goldprice.org/cryptocurrency-price.
[7] https://finance.yahoo.com/lookup.
[8] https://github.com/pushshift/api.
[9] https://www.reddit.com/r/CryptoCurrency.

2.3 Google News Titles

Another source that provides cryptocurrency statements is the Google news API[10]. This API serves the 100 most recent news titles that include the requested keyword. We extract information about 153 $(n*(n-1)/2)$ distinct pairs of cryptocurrencies. For each pair, we receive 100 news from the API which contributes to creating one dataset for each pair. Each one of them contains a title, link to the source, as well as date of the news publication.

3 Methodology

We construct seven networks of cryptocurrencies using seven different approaches, aiming to express the interrelationships between the selected cryptocurrencies. The wide range of relations between the cryptocurrencies, encoded in the networks, allows us to gain insights into the cryptocurrency market dynamics, while observing how they affect each other. Each of these seven networks, their creation, their significance and their analysis are explained in depth in [2], but regardless, we are introducing them in this paper as well.

3.1 Cryptocurrency Daily Price Correlations

The first relationship that we are exploring is the one between the cryptocurrencies' prices. We calculate the correlation of their prices which results in our first network.

3.2 Cryptocurrency Daily Return Correlations

Moreover, the daily prices of the cryptocurrencies helps us to calculate the daily returns. Firstly, we divide the price of a cryptocurrency from the present day by the price from the day before. Then we apply a natural logarithm on the computed ratio to get the daily return. The mathematical representation is shown in the following equation.

$$Daily\ Return = \ln\left(\frac{present\ day\ closing\ price}{prior\ day\ closing\ price}\right) \qquad (1)$$

The daily price returns correlations produce the second network.

3.3 Reddit Title Sentiment Correlations

Sentiment analysis represents a methodology for extracting emotions from text that helps to establish the author's perspective towards a subject. Thus, we can identify if the author's attitude is mainly positive or negative. To evaluate the Reddit titles, we use a pre-trained model for sentiment analysis [3] that is based

---

[10] https://news.google.com/rss/search?q=Bitcoin+Etheriumhl=en-USgl=USceid=US:en.

on RoBERTa transformer architecture [4]. Additionally, we fine-tune this model on short financial statements. Our aim is to compute the news sentiments for each cryptocurrency, on a daily basis, using the collected titles from Reddit groups. Whenever there are multiple news in one day for a cryptocurrency, we calculate the average of all news sentiment as daily sentiment for that cryptocurrency. Due to this, we have one news sentiment time series per cryptocurrency.

We assign weights to the pairs of cryptocurrencies in the network by calculating the correlations between cryptocurrency sentiments.

3.4   Co-occurrences of Cryptocurrencies in Google News

To compute the daily occurrence frequency of the cryptocurrency pairs, we utilize the Google news API. For each pair, we send a query to the API which returns back a total of 100 news articles. Using this data, we determine the average daily frequency by dividing the amount of news associated to a cryptocurrency pair (100) to the total days that passed between the first and last news.

$$Average\ Daily\ Frequency = \frac{number\ of\ news\ items\ for\ pair}{time\ period\ between\ the\ first\ and\ last\ news} \quad (2)$$

Using the calculation from Eq. (2), we obtain the average of news items published each day for a single cryptocurrency pair. The result represents the significance of the relationship between cryptocurrency pairs and it serves in creation of the fourth network.

3.5   Co-occurrence of Cryptocurrencies in Reddit Groups

We follow the same approach described in Sect. 3.4 to extract the interrelationships from Reddit headlines. For each of the 153 cryptocurrency pairs, we query the Reddit API and search for titles that include the names of both cryptocurrencies which constitute the pair. We obtain the frequency of cryptocurrency pair occurrence using the same formula specified in Eq. (2). We utilize these frequencies in constructing the fifth network.

3.6   Explainable ML for Cryptocurrency Price Forecasting

Explainable AI refers to techniques that enables humans to understand and comprehend the ML model's results. To simplify, AI model takes inputs and produce output, thus Explainable AI models presents the impact that inputs have on the output. As explainable model, we choose the SHAP (SHapley Additive exPlanations) [5] which helps us evaluate the XGBoost machine learning model [6]. We use the XGBoost model to forecast the cryptocurrencies' prices. As input to the XGBoost model, we provide the historical prices and sentiments of the cryptocurrencies. Afterwards, we utilize the SHAP model for building a new network on the basis of the cryptocurrencies' relationships. We use the combination of these two models since their bond results in a successful story, as shown in previous research [7–9].

The XGBoost model is trained with 80% of the input data, while the remaining 20% are used for testing. We then use SHAP to explain the results from the XGBoost model. The SHAP model clarifies how input features influence the forecasting model. As output, SHAP provides impact value for each cryptocurrency when predicting a particular cryptocurrency's price. These relation values helps us to create the sixth network.

3.7 Explainable ML for Cryptocurrency Daily Return Forecasting

Additionally, we repeat the method explained in Sect. 3.6, this time we utilize the Explainable AI model in relation to the cryptocurrency price returns. It's important to note that both networks, created using the Explainable ML approach, have directed links in the network due to the nature of the SHAP process, i.e., one cryptocurrency's influence over another may not be the same in the opposite direction.

4 Results

We try to find the currency which is the most dominant and impacts the prices to rise and fall. To achieve this, we apply centrality measures such as Eigenvector Centrality and Closeness Centrality on our networks. The aim is to observe the nodes', i.e. cryptocurrencies' reaction when a change in the price of a singular cryptocurrency occurs. The result from that observation is extracting hidden information that will help us identify the cryptocurrencies that are valuable to invest in when such change in the price occurs. In addition to this, centrality measures help us to spot the currency that is closest to all other currencies, in order to know which one of them is in the best position to affect the entire network most quickly. This will allow investors to speedily react in favor of preventing loss or gaining money by investing in the affected cryptocurrencies.

To obtain the results, we use NetworkX, a Python library for studying and manipulating networks. The only exception is the Node Strength measure that we calculate using our function. This function calculates the Node Strength as a simple sum of the links' weights from/to a certain node.

Following are five centrality measures that we use to predict the network's behavior.

4.1 Degree Centrality

Degree Centrality is simply the number of links each node holds, i.e. a count of how many edges a node has. This allows us to find the most central, most popular nodes. Degree Centrality normalizes the values by dividing by the maximum possible degree in the network. Hence, the range of each value is from 0 to 1.

However, the networks obtained using the Explainable ML approach are represented as a directed graph, thus the number of a node's links can be doubled in comparison to undirected graphs. Therefore the highest value is 2. We present the top 5 values of this centrality measure below, for each of the seven networks.

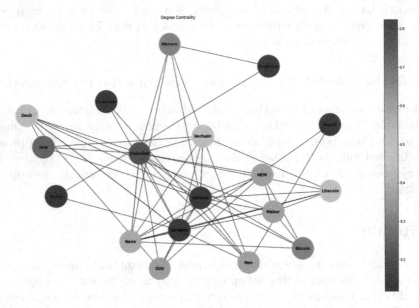

Fig. 1. Degree Centrality measured on explainable ML for cryptocurrency price forecasting network

- Cryptocurrency daily price correlations - Ethereum(0.43), Nano(0.43), VeChain(0.43), Maker(0.43), Cardano(0.43)
- Cryptocurrency daily return correlations - EOS(0.75), Ethereum(0.75), Litecoin(0.67), Neo(0.5), Monero(0.5)
- Cryptocurrency Reddit titles sentiment correlations - Dash(0.41), Litecoin(0.35), Ethereum(0.35), Cardano(0.29), NEM(0.24)
- Co-occurrences of cryptocurrencies in Google news - Dogecoin(0.86), Bitcoin(0.64), Ethereum(0.43), Cardano(0.36), VeChain(0.36)
- Co-occurrence of cryptocurrencies in Reddit groups - Bitcoin(0.92), Ethereum(0.62), EOS(0.54), Ripple(0.46), Cardano(0.46)
- Explainable ML for cryptocurrency price forecasting - Cardano(0.8235), Celsius(0.8235), Chainlink(0.7647), Nano(0.5882), VeChain(0.5294)
- Explainable ML for cryptocurrency daily return forecasting - Celsius(0.9412), EOS(0.9412), Dash(0.6471), Chainlink(0.5294), Nano(0.4706)

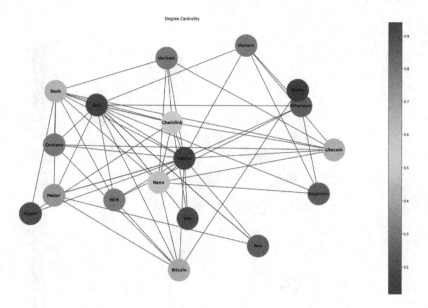

Fig. 2. Degree Centrality measured on explainable ML for cryptocurrency daily return forecasting network

4.2 Eigenvector Centrality

Similarly to Degree Centrality, Eigenvector Centrality also takes into consideration a node's overall number of links. Additionally, this centrality measure goes deeper by paying regard to the connectivity of the neighbors' nodes, calculating the number of their connections, and so forth throughout the network. This approach reveals which nodes have the broadest reach in the entire network.

The highest score for Eigenvector Centrality measure for our networks is 0.68, while the lowest is $-3.2e-16$. Below are the top 5 values of this measure for each network.

- Cryptocurrency daily price correlations - Ethereum(0.39), Cardano(0.38), Nano(0.38), Maker(0.36), VeChain(0.31)
- Cryptocurrency daily return correlations - Ethereum(0.41), EOS(0.41), Litecoin(0.4), Neo(0.34), Monero(0.31)
- Cryptocurrency Reddit titles sentiment correlations - Litecoin(0.48), Dash(0.36), Maker(0.36), Ripple(0.33), Cardano(0.31)
- Co-occurrences of cryptocurrencies in Google news - Bitcoin(0.59), Dogecoin(0.49), Ethereum(0.45), Cardano(0.32), VeChain(0.23)
- Co-occurrence of cryptocurrencies in Reddit groups - Bitcoin(0.68), Ethereum(0.66), Litecoin(0.2), Ripple(0.15), EOS(0.13)
- Explainable ML for cryptocurrency price forecasting - Cardano(0.66), Litecoin(0.6307), Dash(0.3059), EOS(0.2624), Iota(0.0647)
- Explainable ML for cryptocurrency daily return forecasting - EOS(0.6711), Dash(0.444), Bitcoin(0.3472), Litecoin(0.3175), Neo(0.2115)

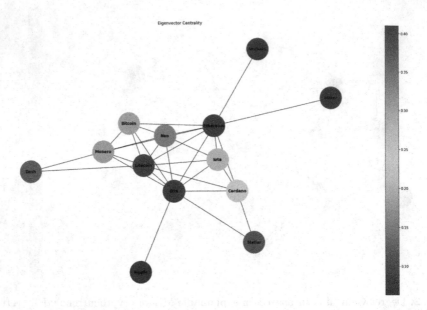

Fig. 3. Eigenvector Centrality measured on cryptocurrency daily return correlations network

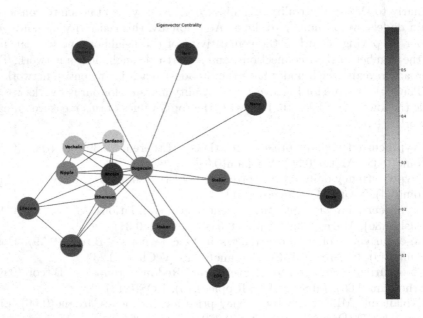

Fig. 4. Eigenvector Centrality measured on co-occurrences of cryptocurrencies in Google news network

4.3   Node Strength

Node Strength, as the name says, expresses the strength of relations or links that a node has with his neighbors. By summing up the links' weights that originate from or end in a certain node, we get the node strength. Because of that, the higher the value, the stronger node's relations are.

The highest score for Node Strength measure for our networks is 14.96, while the lowest is 0.0036. Following, we present the top 5 values of this measure for each network.

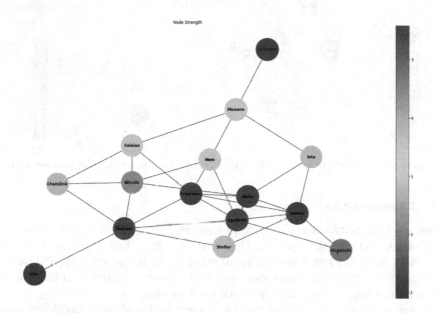

Fig. 5. Node Strength measured on cryptocurrency daily price correlations network

- Cryptocurrency daily price correlations - Nano(5.57), Maker(5.55), VeChain(5.51), Ethereum(5.5), Cardano(5.47)
- Cryptocurrency daily return correlations - Ethereum(7.61), EOS(7.46), Litecoin(6.71), Neo(4.95), Monero(4.87)
- Cryptocurrency Reddit titles sentiment correlations - Litecoin(0.56), Dash(0.54), Ethereum(0.5), NEM(0.38), Maker(0.38)
- Co-occurrences of cryptocurrencies in Google news - Bitcoin(14.96), Dogecoin(12.5), Ethereum(8.1), Cardano(5.23), VeChain(4.08)
- Co-occurrence of cryptocurrencies in Reddit groups - Bitcoin(4.26), Ethereum(3.29), Litecoin(0.92), Ripple(0.8), EOS(0.75)
- Explainable ML for cryptocurrency price forecasting - Celsius(0.0422), Cardano(0.0417), Chainlink(0.0276), Nano(0.026), Litecoin(0.0244)
- Explainable ML for cryptocurrency daily return forecasting - EOS(0.0589), Celsius(0.0572), Chainlink(0.0317), Dash(0.0284), Nano(0.0254)

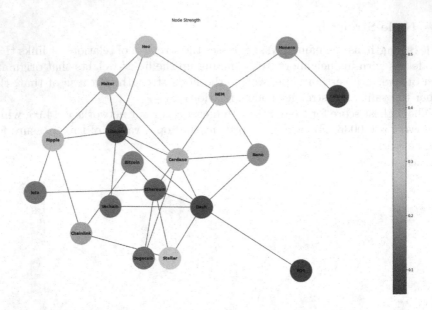

Fig. 6. Node Strength measured on Reddit title sentiment correlations network

4.4   Closeness Centrality

Closeness Centrality assigns a score to each node that signifies how 'close' it is to all other nodes in the network. Hence, those with a high value for Closeness Centrality have shortest distances to all other nodes in the network. This allows us to find the cryptocurrencies that are in the greatest position to influence or to pass a message to the entire network most quickly.

Because this approach calculates the shortest path and our network links have weights representing a relation score, we alter the values to represent distance instead of relation. To do this, we find the difference between the maximum link weight and the second highest link weight. We then transform each of the weights by calculating the subtraction between the maximum link weight and a certain link weight. On top of that we add the previously calculated difference between the maximum link weight and the second highest link weight in order to avoid a weight of 0. Therefore, the highest relation scores are now the lowest distance weights in the network.

We use this data to calculate Closeness Centrality, Betweenness Centrality and to visualize the output.

The highest score for Closeness Centrality measure for our networks is 202.8591, while the lowest is 0. We show the top 5 values of this measure for each network in the following section.

- Cryptocurrency daily price correlations - Ethereum(14.42), Nano(13.39), Celsius(13.26), Stellar(13.19), Maker(13.18)
- Cryptocurrency daily return correlations - Ethereum(15.66), Litecoin(15.56), EOS(14.84), Bitcoin(13.56), Cardano(9.96)

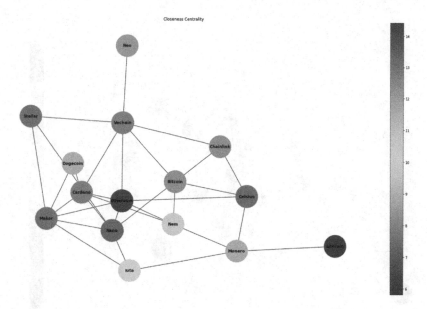

Fig. 7. Closeness Centrality measured on cryptocurrency daily price correlations network

- Cryptocurrency Reddit titles sentiment correlations - Ethereum(5.28), Dash(5.27), Litecoin(5.04), Stellar(4.82), NEM(4.79)
- Co-occurrences of cryptocurrencies in Google news - Dogecoin(0.44), Bitcoin(0.43), Ethereum(0.43), Cardano(0.43), VeChain(0.32)
- Co-occurrence of cryptocurrencies in Reddit groups - Bitcoin(0.25), Ethereum(0.21), EOS(0.17), Ripple(0.17), Cardano(0.16)
- Explainable ML for cryptocurrency price forecasting - Litecoin(202.8591), Cardano(194.9919), EOS(152.7164), Dash(119.2337), Iota(99.7176)
- Explainable ML for cryptocurrency daily return forecasting - EOS(124.3702), Bitcoin(91.0175), Dash(85.6665), Litecoin(75.1435), Neo(65.18)

4.5   Betweenness Centrality

Betweenness Centrality expresses how much a given node is in-between others i.e. the total occurrences of a node on the shortest path between other nodes. This information allows us to find the nodes that control the information flow in the system.

We apply the same methodology as in Closeness Centrality to alter the weights in our networks as ones that represent distance. The highest score for Betweenness Centrality measure for our networks is 0.81, while the lowest is 0. Following are the top 5 values of this measure for each of the seven networks.

- Cryptocurrency daily price correlations - Nano(0.29), VeChain(0.27), Ethereum(0.18), Maker(0.14), Monero(0.14)

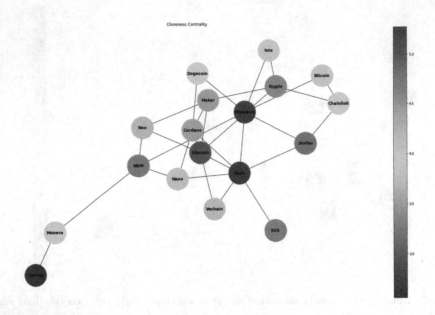

Fig. 8. Closeness Centrality measured on Reddit title sentiment correlations network

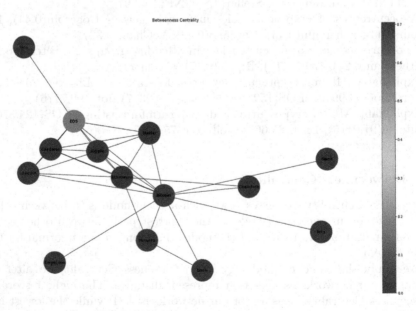

Fig. 9. Betweenness Centrality measured on co-occurrence of cryptocurrencies in Reddit groups network

- Cryptocurrency daily return correlations - Ethereum(0.61), Litecoin(0.41), EOS(0.3), Bitcoin(0.0), Cardano(0.0)
- Cryptocurrency Reddit titles sentiment correlations - Ethereum(0.3), NEM(0.28), Dash(0.26), Litecoin(0.21), Monero(0.12)
- Co-occurrences of cryptocurrencies in Google news - Dogecoin(0.81), Bitcoin(0.25), Stellar(0.14), Ethereum(0.1), Cardano(0.0)
- Co-occurrence of cryptocurrencies in Reddit groups - Bitcoin(0.71), EOS(0.15), Ethereum(0.0), Litecoin(0.0), Ripple(0.0)
- Explainable ML for cryptocurrency price forecasting - Chainlink(0.1029), Cardano(0.0956), NEM(0.0551), Neo(0.0368), Nano(0.0331)
- Explainable ML for cryptocurrency daily return forecasting - EOS(0.5919), Maker(0.2316), Celsius(0.2243), Chainlink(0.2132), Nano(0.0809)

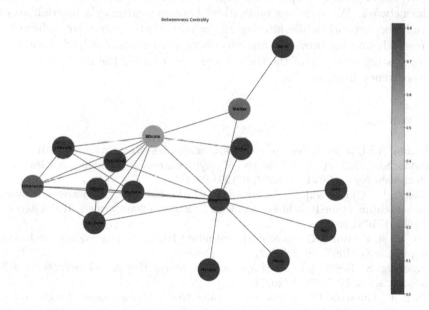

Fig. 10. Betweenness Centrality measured on co-occurrences of cryptocurrencies in Google news network

The results from all five centrality measures that we employ in this paper, shows that Ethereum is one of the most central cryptocurrency. Ethereum is a leader in the number of appearances in the top 5 cryptocurrencies for each network and each centrality measure. It appears 24 times out of a possible 35 times. Right after are Cardano with 19, Litecoin with 18, EOS with 17 and Bitcoin with 14 appearances out of a total 35. On the other hand, the least central cryptocurrencies are Stellar and Iota with less than 5 appearances.

5 Discussion and Conclusion

This paper provides a methodology for creating seven distinct cryptocurrency networks as well as detecting the most influential currency in the network. We also search for the one that project its impact rapidest onto others. Moreover, we look into the effect that social media have over the cryptocurrency market.

We study the relations between the cryptocurrencies from different perspectives. Starting with correlation between their prices, as well as their returns, we continue with exploring the social media influence by employing sentiment analysis techniques considered state-of-the-art. Next are occurrence frequencies of cryptocurrency pairs in Google news and Reddit, followed by explainable AI model that provides clarification of how one cryptocurrency affects the others. Finally, we calculate centrality measures on each of these networks with the purpose of gaining deeper knowledge of the relationship between currencies in the network. We strive to understand how one currency's reaction spread through the network while altering the others. Furthermore, we believe that our research could improve the analysis of cryptocurrencies' behavior and boost the forecasting accuracy of the their prices, i.e. reducing the uncertainty in the cryptocurrency market.

References

1. Satoshi, N.: Bitcoin: A peer-to-peer electronic cash system. Manubot (2019)
2. Todorovska, A., et al.: Analysis of cryptocurrency interdependencies. Proc. Blockchain Kyoto **2021**(BCK21), 011004 (2021)
3. Mishev, K., Gjorgjevikj, A., Vodenska, I., Chitkushev, L.T., Trajanov, D.: Evaluation of sentiment analysis in finance: from lexicons to transformers. IEEE Access **8**, 131662–131682 (2020)
4. Liu, Y., et al.: RoBERTa: a robustly optimized BERT pretraining approach. arXiv preprint arXiv:1907.11692 (2019)
5. Lundberg, S., Lee, S.: A unified approach to interpreting model predictions. arXiv preprint arXiv:1705.07874 (2017)
6. Chen, T., Guestrin, C.: XGBoost: a scalable tree boosting system. In: Proceedings of the 22nd ACM SIGKDD International Conference on Knowledge Discovery and Data Mining, pp. 785–794 (2016)
7. Parsa, A.B., Movahedi, A., Taghipour, H., Derrible, S., Mohammadian, A.K.: Toward safer highways, application of XGBoost and SHAP for real-time accident detection and feature analysis. Accid. Anal. Prev. **136**, 105405 (2020)
8. Bi, Y., Xiang, D., Ge, Z., Li, F., Jia, C., Song, J.: An interpretable prediction model for identifying N7-methylguanosine sites based on XGBoost and SHAP. Mol. Ther. Nucleic Acids **23**, 362–372 (2020)
9. Meng, Y., Yang, N., Qian, Z., Zhang, G.: What makes an online review more helpful: an interpretation framework using XGBoost and SHAP values. J. Theor. Appl. Electron. Commer. Res. **16**(3), 466–490 (2021)

Applied Artificial Intelligence

Evaluating Micro Frontend Approaches for Code Reusability

Emilija Stefanovska and Vladimir Trajkovik[✉]

Faculty of Computer Science and Engineering, Skopje, North Macedonia
emilija.stefanovska@students.finki.ukim.mk,
trvlado@finki.ukim.mk

Abstract. The term micro frontend is relatively new, and it is a continuation of the microservice architecture on the client-side. This paper researches the possibility to use micro frontend architecture for frontend code reusability. The goal of this paper is to achieve code reusability by providing a good organizational structure by using a micro frontend approach. Reusing the code should help to increase the time to market and allow for better scalability. The paper starts by overviewing the base characteristics of domain-driven design as one of the key concepts behind micro frontend architecture. It continues with evaluating existing micro frontend architectures by a set of defined qualitative attributes. Lastly, it uses the findings from the evaluation in building a technical solution. The implementation process can be divided into two phases: decomposition of an existing frontend application; reusing the code by integrating the decomposed components in a new micro frontend application. The results of the implementation process should confirm the micro frontend approach for code reusability.

Keywords: Micro frontend · Code reusability · Domain-driven design · Web components · Module Federation

1 Introduction

From the appearance of the internet in 1990 until today, web technologies are constantly developing. With the increased usage of websites, the complexity of backend applications also increased [1]. Continuously adding new functionalities increases the complexity and size of the application and slows down the testing and delivery processes. Such an application becomes difficult to maintain or change, but it also becomes difficult to understand its domain logic [2]. One solution to these problems is to split the backend application into smaller independent parts that will be developed and delivered independently and together will function as one application called micro-services [3]. One of the main challenges of the micro-service architecture is how to divide the application into smaller independent parts. One way to split the application is with the help of a discipline called domain-driven design (DDD). DDD is one of the base concepts behind micro-service architecture. It provides methods and tools for structuring applications with a complex domain and can help in defining the microservices [4, 5].

Disadvantages faced by large monolithic backend applications also apply to client-side applications. Today many organizations are facing the problem of unmaintainable code on the client-side. The size of the code base doesn't allow to easily add new functionalities, change technology, or the development framework to follow the fast-growing technology trend. Many organizations come to a point where they want to replace their existing client-side application with a new, modern, and faster application to be more concurrent on the market. Rewriting the application's code is one potential solution to the problem but for many organizations, this proved to be a strategic mistake [6, 7]. Another solution to the problem is to continue the split of the backend application to the client-side application and this is the main idea behind the micro frontend architecture [8]. Each micro frontend application can be responsible for a single business domain and can be developed, tested, and delivered independently from other micro frontend applications. This organization should keep the code bases small so it's easier to understand the domain logic [9]. Hence, this research will explore the possibility of using a micro frontend approach for achieving code reusability. The goal of this paper is to achieve code reusability by providing a good organizational structure with micro frontend architecture. Code reusability has the potential of speeding the time to market but also providing modularity and scalability in the new application.

This research paper consists of six sections. The first section is the introduction. It gives a historical overview of the research problem and defines the goal of the paper. The second section of the paper investigates domain-driven design. The third section should serve as a technical background for understanding the technical decisions in the implementation process. It gives an overview of the micro frontend architecture and concludes by evaluating the different micro frontend composition types. The fourth section is the practical part of this paper. The implementation process is split into two phases. The first phase looks at decomposition approaches while the second phase provides two use cases for building a micro frontend application. The fifth section presents the results of a questionnaire conducted among software engineers experienced in micro frontend development. The sixth last section summarizes the results of the implementation process.

2 Domain-Driven Design

Every software is designed and developed to execute some activity that is of interest to its users. The area of this activity is the domain of the system. To build any software, we must first understand its domain and if this domain is complex the amount of information can become huge [10]. To better maintain a complex system, it's best to divide it into smaller, less complex, and well-defined logical components. The components should be decoupled from each other, and the dependencies should be kept to a minimum [11]. This should make the navigation through the codebase easier for the developers. For this reason, the process of adding new features or resolving bugs should be faster and more efficient. Domain-driven design is a discipline that gives directions on how to structure and divide a complex system into smaller components.

Domain-driven design can be separated into two disciplines: model-driven design and strategic-driven design [11]. In some literature, the model-driven design is also called tactical-driven design, but this term does not originally come from E. Evan's literature. The model-driven design is oriented towards code implementation and gives directions on how to better structure the code. Some examples include using layered architecture, services, and entities. On the other hand, strategic-driven design is oriented toward the architecture of complex applications. The main goal of the strategic-driven design is to identify and divide the domain of the application into subdomains. To define the subdomains, there are three base concepts we must understand first:

- Bounded context. Represents a logical boundary between two domains, and it is intended to hide the implementation details between them.
- Ubiquitous language. Should be used by both technical and domain experts as a common language that connects everyday communication to the code implementation.
- Context map. A context map gives a picture of all the bounded contexts that exist in a system. It also shows the points of interaction between two bounded contexts [10].

Both modal-driven and strategic-driven design will be of further interest to this paper. The base of the strategic-driven design was to define the subdomains of the system. To define the subdomains, we must first look at the processes the system executes. As an example of defining subdomains, we implemented a prototype application that represents a simplified version of a banking system. It executes two simple processes which will be referred to as P1 and P2. Let's say process P1 is connected to all operations involved with banking cards while P2 is with bank offices. The simplest heuristic for defining the sub-domains in this example is vocabulary. Now if we consider the vocabulary, we can identify two subdomains which will be referred to as D1 and D2. Now each of these subdomains can be modeled separately.

The application is implemented as a standard web application consisting of a back-end part implemented in Java Spring Boot and a single-page client application implemented in Angular. Since this research is more oriented towards frontend architecture and development, we will focus on the client application. It's important to note that not all domain-driven principles are applicable to client applications although some of them have a fundamental meaning. The client application already follows some of the principles and implements processes P1 and P2 as two bounded modules. The next few chapters will research different micro frontend architectures and investigate how to organize and structure the code from the two subdomains and reuse it in a new micro frontend application.

3 Background

3.1 Micro Frontends

Micro frontend architecture is an architecture for building client-side applications. This architecture suggests dividing the client-side application into smaller front-end applications that together work as one application for the end-user.

Before starting a micro frontend project there are several decisions that need to be made. These decisions will determine the future course of the project. In his literature, Mezzalira puts these decisions in a so-called micro frontend decision framework which consists of four parts: definition of micro frontend, composition, routing, and communication [12].

The first part is about defining what a micro frontend is in the context of the application being built. The application can show multiple micro frontends on one screen or one micro frontend per screen. With this definition, Mezzalira divides the micro frontend architecture into two parts: horizontal and vertical micro frontend architecture. A horizontal micro frontend application allows multiple micro frontends on the same screen. This indicates that two or more teams can be responsible for one screen. This organization requires more coordination between the teams to have consistent design decisions. The micro frontends might also require communication to share information about the user interaction. A vertical micro frontend application shows one application per screen. This indicates that every team is responsible for one business domain. The vertical micro frontend application is tightly coupled with the domain-driven design that can be used in defining the micro frontends.

The second part of the framework refers to the integration of the micro frontends into one composition. In the literature [9, 12–14], there are three types of compositions: server-side, edge-side, and client-side composition.

The third part of the framework is about routing between the micro frontends, and it's tightly coupled to the composition type. If the application is composed on the server side the routing must be done on the server side. As opposed to this if the application is composed on the client side the application routing will be done by the client application.

The last part of the framework is about the communication between the micro frontends. In an ideal case, the micro frontend applications should be completely independent and wouldn't have a need to communicate with each other. This is especially true for the vertical micro frontend architecture. Some possible communication solutions include custom events, web storage, and query strings [12].

The micro frontend architecture comes with many benefits like small code bases, autonomous teams, scalable code, and fault isolation [9, 12]. On the other hand, it introduces other issues like code redundancy, routing, and communication between the applications. Because of this, not every application is suitable to follow the micro frontend approach.

3.2 Micro Frontend Composition Types

As previously mentioned, there are three micro frontend composition types. According to Mezzalira L., only the client-side composition is suitable for a vertical micro frontend because it gives the closest experience to a SPA. All other three composition types can be used for the horizontal micro frontends [12]. Authors Jackson K. and Geers M. additionally divide the client-side composition into build-time and runtime client-side composition [9, 13].

This section will give an overview of all three composition types with some of their benefits and common issues and will serve as a background for understanding the technical decisions made during the implementation process.

Server-Side Composition

The composition of the final HTML page is done by a server that can be a simple proxy server like Nginx or a custom application with additional logic. This server is located between the micro frontend application servers and the client or ideally a CDN [12]. The main benefit of this approach is the performance. The server can use caching and additional logic to decrease the calls to the micro frontend servers which can result in a short loading time of the pages. Ideally, all servers would be in the same data center where the network latency is much smaller. The biggest issue with this approach is the client experience when interacting with the application. If this composition is not combined with client-side rendering, then every interaction will require a full page reload [12, 13].

Edge-Side Composition

Edge-side composition is enabled by the Edge side integration language (ESI). ESI is a specification defined by Oracle as one of the co-authors and allows composing an HTML page from HTML fragments [15]. The integration is done on the CDN level like Akamai or AWS Lambda [16]. The content delivery networks allow one resource to be found in multiple locations and assure that the requested resource will be served from the point that's closest to the user. This results in smaller network latency and hence faster load time for the end-user.

Like the server-side composition, this composition also lacks a good user experience if not combined with a client-side composition.

Client-Side Composition

With this type of composition, the final HTML page is composed directly in the client's browser. A micro frontend integrated on the client-side usually consists of several micro frontends and a container application. The container application is responsible for selecting the correct micro frontend and often contains the common application parts like navigation and page footers. The concept is shown in Fig. 1.

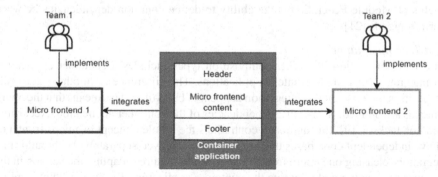

Fig. 1. Client-side composition

Because of the popularity of frontend frameworks like React, Angular, and Vue, the client-side composition is one of the most frequently used micro frontend composition types.

Iframes
Iframes can be defined as a special HTML element that allows the loading of a second HTML document on the same HTML page [17]. Iframes can be considered as one of the oldest ways of building micro frontends. They provide a high level of isolation and the JavaScript and CSS code inside the iframe cannot be affected by any changes in the container application. However, loading many iframes on the same page in the browser can cause serious performance issues [13].

Web Components
Web components represent a set of APIs which allow the building of custom and reusable HTML elements. Today, almost every modern frontend framework like React and Angular allows the usage and creation of web components. There are two main technologies that allow building web components: custom elements and shadow DOM. Custom elements allow the definition of custom HTML elements while the shadow DOM provides the encapsulation of web components [18].

Isolation is one of the main benefits of web components. It allows the integration of micro frontend applications developed in different frontend frameworks. Because of this fact they are a good choice for code reusability. The old code can be encapsulated inside a web component and integrated into a new application.

Module Federation
Module Federation is a relatively new technology initially released in 2020 [19]. It is a plugin of Webpack (a tool designed to bundle JavaScript applications). It allows async loading of JavaScript bundles at runtime so the application can be built by remote and independent modules [20]. The plugin distinguished two main application concepts called remote and host. The host is the container application that integrates the remote applications at runtime while the remotes are the micro frontends that export their application code to the host.

Module Federation as described by the author was built to solve the many problems that micro frontends are facing like routing or common dependencies. One of the major benefits of Module Federation is the ability to define common dependencies between micro frontends [21].

Build Time Integration
Build time integration like other composition types includes a container application that integrates other micro frontend applications. The difference from other client-side composition types is that this integration happens at build time. This means that the micro frontends are defined as any other dependencies of the container application. According to author Jackson [9], this approach contradicts the whole concept of micro frontends to have independent code bases that are built and deployed separately. With build time integration releasing one micro frontend application requires adapting the version in the container application and releasing the container application. Because of this coupling at build time, it might be easier to have the micro frontends as separate modules as it will decrease the complexity of setting up and maintaining multiple applications.

3.3   Evaluation of Micro Frontend Composition Types

This section evaluates the micro frontend composition types bases on the literature findings. To evaluate any software architecture there are several standardized quality attributes to be considered. At the same time, these quality attributes should support the business goals of the project [22, 23]. One major business goal when reusing the old code is to have a faster time to market. For this reason, we can consider the simplicity of the architecture and the development experience as key factors [24, 25]. The old code can be complex and hard to understand. During its lifetime it can be that many people even teams changed. To reuse the old code, it is important that the architecture is simple so that the team working on the old application doesn't have to make a lot of changes to the code base. Simplicity and experience with technology also support the development experience (Table 1).

Table 1. Evaluation of micro frontend composition types.

Qualitative attribute	Micro frontend composition types
Performance	Server-side, web components, Module Federation, build-time
Modularity	Server-side, edge-side, iframes, web components, Module Federation
Testability	iframes, web components, Module Federation, build-time
Developer experience (DX)	Web components, Module Federation, build-time
Simplicity	iframes, web components, Module Federation, build-time
Scalability	Server-side, edge-side, web components, Module Federation

Unlike other composition types, web components and Module Federation are two technologies which fulfill all qualitative attributes. For this reason, they both will be considered when building the technical solution.

4   Implementation Process

The implementation process can be divided into two phases. The first phase includes splitting the code from the existing single-page application into micro frontends. The goal of the decomposition is to provide an organizational structure that will enable us to reuse the functionalities of the application into a new micro frontend application. In this phase, we will investigate two decomposition approaches and compare them.

The second phase of the implementation process uses the result from the comparison in the first phase and integrates the decomposed components into a container application. In this phase, we will provide two use cases of code reusability with web components and Module Federation.

Both phases of the implementation process can be started in parallel. For example, while one team works on the decomposition of the old application a new team can start with building the new micro frontend application and extending it with new functionalities.

4.1 Decomposition Phase

Simulated Decomposition Using Web Components

One way of decomposing the application is to virtually split the independent domain parts of the application with the help of web components. With this type of decomposition, each of the domain modules will only simulate an independent application. We will still use the original SPA and its structure will remain the same.

Every standard Angular SPA has one root component which is the first rendered component when the application launches. This component usually contains a router-outlet directive that tells the Angular router where to render the content of the component for the given route [26]. Because the idea of this simulation is to have every domain module behaving as an independent application, every module will have its own root or entry component which will be used to render the content of the module's components. This component must be defined as a web component so it can be recognized by the browsers as an HTML tag. Then, as soon as the web component is used, the content from the corresponding domain will be rendered. Figure 2 shows the structure of the application before and after the decomposition.

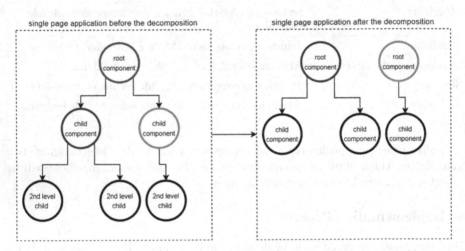

Fig. 2. Structure of the single page application before and after the decomposition.

One important issue to address in this approach is application routing. Since there is an entry web component for each module now, the root Angular component was removed. This means that initially when the micro frontend application is launched no component will be rendered by default. Even more, the end-user will initially interact with the container application. Until the container loads the selected micro frontend its router won't be initialized and therefore register changes. This is the reason why we needed to introduce programmatic routing. The routing by the micro frontend will be done when the parent web component is created.

Decomposition to Separate Applications Using Web Components
The second way of decomposition is to split the domain modules into two independent applications which will be built, tested, and deployed independently. The simplicity of this decoupling largely depends on the current application implementation. If there is a tight coupling between the modules the split can be almost impossible since many components would need to be reworked or even rewritten. However, if the application follows DDD principles and every domain is bounded to its own module the split can be done relatively easily and each module can be moved to a separate application. Figure 3 illustrates the decomposition of a SPA into two separate applications.

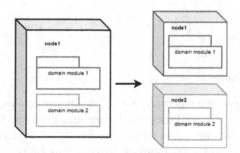

Fig. 3. Splitting the domain packages of an application into separate running applications.

The first step of this decomposition would be to create a project for each module. Now each application would need to install its own dependencies to work properly. Depending on the composition type that will be used there are certain changes that need to be made to each application. If the web component composition type is used, then the root component of each application should be transformed as a web component in the same way as it was described in the previous section.

In the previous decomposition approach, the web component class used programmatic routing. This decomposition approach also needs to implement the routing programmatically. The main reason is the control over each micro frontend router in the container application. The second reason is deep links. All further details will be explained in use case 1 of the integration phase.

Comparison
Table 2 compares the two decomposition approaches based on some of the fundamental micro frontend principles.

The table clearly shows that the second decomposition approach has all the benefits of the micro frontend architecture and for this it will be used for building the new micro frontend application.

Table 2. Comparison of decomposition approaches

Micro frontend principles	Simulated decomposition using web components	Decomposition to separate applications
Independent delivery	False	True
Independent technologies	False	True
Autonomous teams	False	True
Autonomous code bases	False	True
Organized around business capabilities	True	True
Scalability	True	True

4.2   Integration Phase

This section will present the integration phase of the implementation process. This phase includes building a container application and integrating the micro frontend applications we introduced in the decomposition phase. The result of this phase should confirm or deny the suggested organizational approach for code reusability.

Use Case 1
The first use case for composing a micro frontend will use web components as one composition type with most benefits.

Due to the popularity of the development framework and our experience with it, the container application will be implemented as a single-page application in Angular. For navigation, the application will use the built-in Angular router. The router matches the browser's URL to a corresponding component, and it expects this component to be a part of the application. Because the corresponding component is in one of the remote micro frontends a helper component had to be implemented. Now every route matches the helper component.

The helper component holds the main logic for selecting the micro frontends. The component holds a configuration map for each micro frontend. The map defines which micro frontend should be loaded for a given path and the server location to loaded from. When the helper component is instantiated, it first checks if the micro frontend is already loaded in the browser by using a special id attribute. If not, it adds the micro frontend bundled application scripts to the HTML page.

This setup of the container application required certain adaptations to the micro frontends. One major problem that occurred with this solution was the routing. The container application is only responsible for the external routing between the micro frontends. It only looks for the first part of the URL to select the correct micro frontend application. After this, the router of the selected micro frontend looks at the full URL to find the correct page inside the micro frontend. Figure 4 illustrates how this works.

When a micro frontend is launched it initializes its own router which listens to changes until the web component is destroyed. Because the container application doesn't get notified when the micro frontend web component is destroyed, in the meantime it

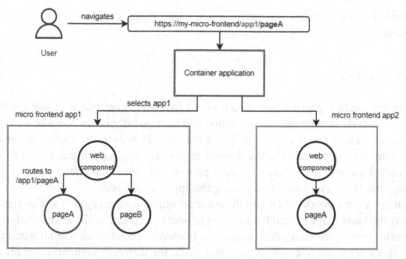

Fig. 4. Routing in the micro frontend application.

can launch a second micro frontend. This causes the router of the first micro frontend to fail with an error because of an unknown route. A solution to the problem was to ensure that only one micro frontend router is active at a time. This can be achieved through conditionally displaying the router-outlet directive which was used to tell the router where to load the content. When the web component is created or when the browsers back and forward buttons are selected, it will check the first part of the URL. If the path is relevant to the micro frontend, it will display the router-outlet directive and render the correct content.

Use Case 2
This use case for composing a micro frontend will use Module Federation as the second composition type with the most benefits.

The container application will be implemented as a single-page application in Angular. One way to integrate Module Federation into the application is to add it as a project dependency. The next step is to define a new webpack.config.js file. This file will contain all the configuration needed to fetch and load the micro frontends. In comparison with the previous solution, the code needed to fetch and load the micro frontend bundles is isolated in one configuration file. Some of the configurations include the resource location of the micro frontends, common dependencies, and versioning of common packages.

The web component composition type we discussed in the previous section required certain adaptations to the micro frontend applications and routing logic was one of them. With this approach, these changes are not needed. This way the application structure remains almost identical to the original application.

For the container application to load the two micro frontends, they should be configured to use Module Federation for exposing the application code. For this reason, Module Federation is added as a project dependency in both micro frontends. The second step would be to configure Module Federation by adding a webpack.config.js file.

The configuration includes the modules which are being exported as well as shared dependencies.

5   Evaluation

The implementation process proves that micro frontends can indeed be used to better organize the code of an existing application and successfully reuse it in a new environment which should result in a faster time to market. However, the application used in the implementation process is a small proof-of-concept application that has a relatively small code base and independent domain parts which made the decomposition of the domain parts faster compared to rewriting the application code.

In this part, we conducted a questionnaire among software engineers. The goal is to compare the knowledge gained from the implementation phase with the knowledge from real-world experience with micro frontends. It's worth mentioning that all participants have experience with micro frontends and work for different companies or projects. The interviewees were asked about the main benefits and downsides of using micro frontends and to provide the use cases where they used a micro frontend application instead of a monolith. Almost all interviewed candidates (80%) answered that the main benefit of using a micro frontend is maintainability due to smaller code bases. The same group thinks that independent deployments and teams are also a major benefit. Having teams that are responsible for their own release cycles and are deploying to production when they are ready, regardless of other teams is a way to achieve faster time to market. However, independent deployments are only possible when the applications are completely independent, or the communication interfaces are not affected by the release changes. Half of the interviewees answered that developing a micro frontend was faster due to the smaller code bases which seems to confirm the initial hypothesis.

One of the most mentioned downsides of micro frontend was implementing communication mechanisms between the micro frontend applications as well as the initial infrastructure setup which can be complex.

From the obtained responses based on real-world experience with micro frontends, it seems that micro frontends provide faster time to market.

6   Conclusion

This paper investigates the possibility of using a micro frontend architecture to achieve code reusability. The research evaluates the possible micro frontend approaches found in the literature and discusses the technologies that can be used. Based on the literature evaluation, as part of the research, a technical solution was implemented. The technical solution shows how micro frontend architecture can be used to better organize the code of an existing single-page frontend application and how the code can be reused in a new environment. The paper provides two practical use cases of code reusability using two different technologies. The first use case uses web components as one technology for building a micro frontend application. It describes how a web component can serve as a shell for encapsulating a whole frontend application. The second use case uses the

Module Federation plugin for building a micro frontend. This use case shows a different approach that handles the integration logic on the configuration level.

Both technical implementations have many benefits. In terms of modularity, both solutions integrate the micro frontends on the client-side at runtime which makes them highly modular. At the same time, both solutions are highly scalable since the container application fetches the micro frontends as static JavaScript files. Based on the results of the implementation process it can be concluded that using a micro frontend approach for code reusability was successful.

One of the main benefits of using a micro frontend approach for code reusability was to provide a faster time to market. This research puts the main emphasis on the implementation process but does not provide any data about the complexity or time needed to set up deployment or testing processes also mentioned in the responses from the questionnaire. Having a decentralized architecture requires having multiple servers, deployment pipelines, and additional monitoring tools. This requires scaling the release and deployment processes to support multiple applications and so it introduces a new type of complexity. Future work can evaluate the available strategies for deploying micro frontends and investigate how the effort impacts the overall time to market.

References

1. Roesler, V., Barrére, E., Willrich, R.: Special Topics in Multimedia, IoT and Web Technologies, 1st edn. Springer, Switzerland (2020). https://doi.org/10.1007/978-3-030-351 02-1
2. Article on monolithic architecture patterns. https://microservices.io/patterns/monolithic.html. Accessed 11 June 2022
3. Farcic, V.: The DevOps 2.0 Toolkit, 1st edn. Leanpub, Victoria (2016)
4. Article on microservices. https://martinfowler.com/articles/microservices.html. Accessed 11 June 2022
5. Newman, S.: Building Microservices, 1st edn. O'Reilly, Sebastopol (2015)
6. Blog post. https://www.joelonsoftware.com/2000/04/06/things-you-should-never-do-part-i. Accessed 13 June 2022
7. Medium article. https://medium.com/@herbcaudill/lessons-from-6-software-rewrite-stories-635e4c8f7c22. Accessed 13 June 2022
8. Micro Frontends Homepage. https://micro-frontends.org. Accessed 13 June 2022
9. Article on micro frontends. https://www.martinfowler.com/articles/micro-frontends.html. Accessed 13 June 2022
10. Evans, E.: Domain-Driven Design: Tackling Complexity in the Heart of Software, 1st edn. Addison Wesley, Boston (2003)
11. Manfred, S.: Enterprise Angular - DDD, Nx Monorepos and Micro Frontends, 4th edn. Lean Pub, Victoria (2020)
12. Mezzalira, L.: Building Micro-Frontends, 1st edn. O'Reilly Media, Sebastopol (2021)
13. Geers, M.: Micro Frontends in Action, 1st edn. Manning Publications, Shelter Island (2020)
14. Rappl, F.: The Art of Micro Frontends, 1st edn. Packt, Birmingham (2021)
15. W3Schoos. https://www.w3.org/TR/esi-lang/. Accessed 17 June 2022
16. The future of micro frontends. https://betterprogramming.pub/the-future-of-micro-frontends-2f527f97d506. Accessed 07 June 2022
17. W3Schoos. https://www.w3schools.com/tags/tag_iframe.asp. Accessed 08 June 2022

18. Web Components Introduction. https://www.webcomponents.org/introduction. Accessed 08 June 2022
19. Micro-frontends building blocks: Webpack Module Federation. https://dev.to/aws-builders/micro-frontends-building-blocks-webpack-module-federation-360a. Accessed 14 June 2022
20. Webpack Module Federation. https://webpack.js.org/concepts/module-federation. Accessed 13 June 2022
21. The Micro frontend Revolution: Module Federation with Angular. https://www.angulararchitects.io/en/aktuelles/the-microfrontend-revolution-part-2-module-federation-with-angular/. Accessed 14 June 2022
22. Carnegie Mellon University Library. https://resources.sei.cmu.edu/asset_files/Webinar/2009_018_101_22232.pdf. Accessed 15 June 2022
23. ISO 25010. https://iso25000.com/index.php/en/iso-25000-standards/iso-25010. Accessed 15 June 2022
24. Medium article. https://medium.datadriveninvestor.com/the-case-for-favoring-simplicity-in-software-49fa9caf8da#:~:text=The%20case%20for%20favoring%20any,to%20the%20most%20important%20first. Accessed 15 June 2022
25. Good Developer Experience Practices. https://developerexperience.io/practices/good-developer-experience. Accessed 15 June 2022
26. Angular Bootstrapping Guide. https://angular.io/guide/bootstrapping. Accessed 15 June 2022

Combining Static and Dynamic Features to Improve Longitudinal Image Retrieval for Alzheimer's Disease

Katarina Trojachanec Dineva$^{(\boxtimes)}$ [iD], Ivan Kitanovski, Ivica Dimitrovski,
Suzana Loshkovska, and for the Alzheimer's Disease Neuroimaging Initiative

Faculty of Computer Science and Engineering, Ss. Cyril and Methodius University, Rugjer
Boshkovik 16, PO Box 393, Skopje, Macedonia
{katarina.trojacanec,ivan.kitanovski,ivica.dimitrovski,
suzana.loshkovska}@finki.ukim.mk

Abstract. The aim of this paper is to enhance medical case retrieval for Alzheimer's disease on the basis of the domain knowledge. We approached the problem in a longitudinal manner, and we represented the medical cases by using different kind of information extracted from Magnetic Resonance Images (MRI) aiming to improve the semantic relevance, precision and efficiency of the retrieval. More particularly, we evaluated the combination of the static, dynamic features and the index reflecting the spatial pattern of abnormality (SPARE-AD) for representing the longitudinal images. According to the obtained results, the combination of the static features representing the volumetric measures along with the cortical thickness measures of the brain structures at the later time point/s together with the dynamic features such as percent change with respect to the value obtained from the linear fit at baseline and symmetrized percent change of the volumetric measures, as well as the index of abnormality provided the best overall retrieval results. The dimensionality of the feature vector was 31–33 features in most of the cases which is significantly lower than in the case of the traditional approach (thousands features in the cases when the whole brain is considered). The approach based on a combination of different kinds of features extracted from the longitudinal data, suggested in this paper, corresponds directly to the nature of the application domain and provides powerful results, yet effective and efficient way for MRI retrieval for AD.

Keywords: Medical cases · Medical images · MRI · Image retrieval · Longitudinal data · Static features · Dynamic features · SPARE-AD · Alzheimer's disease

---

"Data used in preparation of this article were obtained from the Alzheimer's Disease Neuroimaging Initiative (ADNI) database (adni.loni.usc.edu). As such, the investigators within the ADNI contributed to the design and implementation of ADNI and/or provided data but did not participate in analysis or writing of this report (http://adni.loni.usc.edu/wp-content/uploads/how_to_apply/ADNI_Acknowledgement List.pdf)."

1 Introduction

Alzheimer's Disease (AD) accounts for estimated 60–80% of dementia cases [1]. There-fore, it is considered as one of the most common forms of dementia in older adults nowadays [2, 3]. It is a progressive, irreversible, neurodegenerative disease that usually starts as a long asymptomatic stage, called preclinical AD. Then, a symptomatic stage called appears, known as mild cognitive impairment (MCI). This stage may be followed by the third possible phase, dementia, which may vary from mild to severe [2, 4–6].

The early and accurate detection of AD, monitoring the progression of the disease and the condition of the patient in general, finding powerful diagnostic or prognosis biomarkers, detecting patients who are most likely to develop AD, as well as reaction to the therapy, are still open questions. Our research is part of AD-related data organization and usage [7–19]. It provides a good basis for the other research directions related to AD, including (1) defining risk factors and the information usage to develop recommendations for prevention [20, 21], (2) diagnosis and prognosis - finding diagnostic biomarkers and biomarkers for prognosis (biomarkers from cerebrospinal fluid and blood, as well as imaging markers) [10, 20, 22–26], treatment reaction analysis [20, 27, 28]. It should be noted that there is no sharp border between the aforementioned research directions, and they are not mutually exclusive.

In fact, due to the rapid development of the technology and medicine, large amount of data is continuously and rapidly generated as part of the medical cases related to AD. In this regard, blood and cerebrospinal fluid biomarkers, genetic markers, neuroimages, and cognitive tests results, are produced and commonly examined and analysed by the physicians in the process of diagnosis or, slightly wider, in the period of monitoring the patient's condition. Then, they usually remain unused in the clinical databases until the need for their removal. Hence, it is essential to provide a way for efficient storage, organization and representation of these data in an appropriate manner that will enable quick and easy access to the medical cases, as well as clinically relevant, but also, precise and efficient retrieval, providing a way for powerful knowledge discovery from these continuously growing data. The computer science development undoubtedly facilitates and enables information extraction from this kind of data, their analysis, visualization and knowledge discovery. Longitudinal images (images acquired at multiple consecutive time points) play crucial role in this domain and reflect the disease progression. Thus, taking into consideration the progressive nature of the disease, the longitudinal approach to the problem of brain image retrieval for AD is essential to make the retrieval more accurate and clinically relevant.

Several studies focus on brain MRI retrieval for AD. Some of them use the tradi-tional approach for feature extraction and base the descriptors directly on the visual image content. For instance, general-purpose image feature algorithms are evaluated for the purpose of a similar retrieval process in [14]. The main idea is that they might be beneficial because of their simplicity (they consider only one 2D slice). This makes their approach appropriate for large scale systems. Authors in [15] proposed a hybrid features extraction technique. They extracted contrast feature, morphological operated features, and texture-based features from brain MRI. Those features are then hybridized by applying fusion technique. The research in [16] is based on evaluation of texture features extracted by using Gray Level Co-occurrence Matrix algorithm, Law Texture

Energy Measure and a combination of the features extracted with these two methods. Gray-level cooccurrence matrix is also evaluated in [17], together with local binary pattern and colour cooccurrence matrix. On the other hand, authors in [18] used estimated volumes of the brain structures, and cortical thickness measures to generate image descriptors for patients' representation in the multistage classifier. In fact, they incorporate this classifier in the process of AD prediction and retrieval. Finally, with the aim to increase the early detection performance for AD, authors in [19] used pre-trained 3D-autoencoder, 3D Capsule Network, and 3D-Convolutional Neural Network.

Several challenges can be identified considering the existing research related to image retrieval for AD. One important challenge is gaining the semantically relevant retrieval results. Another critical part might be the size of the feature vector, tending to be high dimensional. Despite the power and accuracy, the deep neural networks bring, the difficulty to understand and interpret their way of decision making makes them challenge to be adapted for medical applications [29]. The main challenge that should also be mentioned is whether the problem is approached cross-sectionally or longitudinally. Common limitation in the existing studies is that images are processed and included in the research cross-sectionally, thus loosing potentially crucial information.

Taking this into consideration, our research is aimed to provide better longitudinal image representation that will reflect the early stage of the disease, the patient's condition at one moment (static state of the brain/current brain degeneration) but will also represent the brain changes caused by the disease over time (dynamic aspect/disease progression). For that purpose, this research aims to evaluate a representation containing a combination of static and dynamic features, together with the SPARE-AD index, namely the Spatial Pattern of Abnormality for Recognition of Early Alzheimer's disease. In fact, we suggest using a combination of the best descriptors, investigated and evaluated in the previous research, namely static [11] and dynamic features [12] along with the value of SPARE-AD [13]. Therefore, the different type of information that each of them separately carries, will be utilized and incorporated into one feature vector. For this purpose, we identified four scenarios and then and then we evaluated them to find out if they are complementary and will contribute to better retrieval results.

Providing efficient and precise medical case retrieval based on MRI for Alzheimer's disease may have multiple benefits from different perspective. One is in the direction of decision support in a way of providing clinicians with powerful and relevant information at the right moment by giving the medical cases of other examined patients that are the most similar to the query patient. This is valuable assistance in the clinical environment. The knowledge that can be discovered from the retrieved medical cases, enables a deeper understanding of the disease in general, as well as, specifically, of the condition of the query patient, thus contributing to the completion of the clinical picture of the query patient. This way, the knowledge gained from the retrieval result supports the decision-making process and determination of an adequate treatment/therapy. Pattern discovery and understanding is another important benefit. Retrieved images and cases they belong to, contain valuable information. This information can provide new insights about the disease, biomarkers identification, and exploration of the disease progression. Assessing response to a treatment is another, very important benefit. The information gathered from the other relevant cases, especially if they are treated in a longitudinal manner provide information about the brain changes due to disease. They give insight of the disease progress that might be used to analyze and assess the reaction to a given treatment. Finally, providing a reliable retrieval system is very beneficial for educational purposes. The possibility of searching through the large medical databases is crucial for educational and research purposes. Medical cases of patients with similar structural brain characteristics, anatomical changes or treatment reaction to the query, provide valuable knowledge for students and scientists.

The paper is organized as follows. The materials and methods are covered in Sect. 2. Here we present data used in our study as well as the general approach and evaluation methodology. The experimental results and discussion are presented in Sect. 3. Section 4 contains the concluding remarks.

2   Materials and Methods

2.1   Participants and Inclusion Criteria

Our research is based on the scans and data provided by ADNI (Alzheimer's Disease Neuroimaging Initiative) database [30]. The aim of ADNI is to enable research that will provide an answer to the question whether imaging techniques such as MRI and positron emission tomography (PET), other biological markers, including APOE status, cerebrospinal fluid (CSF) markers, and full-genome genotyping, along with neuro-psychological and clinical assessments, may indicate the presence and allow assessment of the progression of MCI and AD.

For this research, we used the standardized list from ADNI-1, containing images acquired at multiple time points. We selected the subjects that have available scans at four time points (TP 1-4), namely, at baseline, and the 6-, 12-, and 24-month follow-ups and belong to AD or normal control (NL) group. Thus, we obtained a total of 267 patients from the standardized list, 168 in AD group, and 99 in NL group. Patients' demographics information and the timing of scans by clinical group for each time point can be found in [12].

We used this selection criteria because of the following reasons: (1) more time points (ex. 36-month follow-up) is not available for the AD group of patients and, additionally, the total number of patients for whom all the scans are available is reduced by more than 12%; (2) a smaller number of time points would not give enough space for investigation and an opportunity to have a good insight into the problem (3) with the selected time points, we have an opportunity to analyze the problem in the case of equally and unequally spaced available time pointes (depending of which time point is missing).

2.2   Longitudinal Image Retrieval for AD

The image retrieval process consists of generating a representation of the query image and all the images previously stored in the database using the same feature extraction technique. After that, the feature vector (descriptor/representation) of the query image is compared to the feature vector of all other images. All the images in the database are then sorted by similarity to the query image, so that the most similar one is at the top. This sorted list of the database images is the result of the retrieval process. In this research, for a given medical case of a patient for whom MRI was acquired at multiple consecutive time points, the set of images (longitudinal images) is given as a query to the system. The retrieval system provides a sorted list of all other patients in the medical database according to their similarity to the query.

To evaluate the proposed scenarios in this research, we used the standard evaluation metric MAP (Mean Average Precision) for quantitative measurement of the retrieval performance. We also calculated precision at first (top) x returned cases (Px), where x $\in$ {1, 5, 10, 20, 30}, and the precision at first (top) R returned case (RP). In this case, R is the total number of relevant images/cases.

In our research, the retrieved case is considered as relevant if the clinical group (AD/NL) of that patient is the same with the clinical group of the query patient. The precision value is the higher the relevant cases are located higher in the result list. Additionally, we used leave-one-out strategy because of the small number of patients used in the evaluation. This means that the descriptor of each patient was used as a query against all other patients' descriptors stored in the database.

2.3 MRI Processing

MRI processing is a critical and very complex task due to the specific characteristics of neurological images. For this research, we used the fully automated software pipelines from the package FreeSurfer [31]. We selected this software package because it has been shown that its processing and morphometric techniques are robust and stable regardless of the manufacturer of the scanner and the strength of the magnetic field [32, 33]. We approached the problem longitudinally, considering that AD tends to progress over time, and applied the fully automated longitudinal pipeline provided by FreeSurfer. Figure 1 depicts the main steps of the processing flow [33] that we applied to all images in the dataset.

2.4 Representation of Longitudinal Images

After the fully automated longitudinal processing with FreeSurfer, several types of measurements can be calculated for the brain regions and used for analysis. These include volume of the cortical and subcortical regions, cortical thickness, calculated for parts of the cerebral cortex, surface area, etc. (static features). Considering these features, research has shown that the quantitative measurements like volume or cortical thickness are more reliable and superior imaging markers related to AD in comparison with surface area [34–36]. Additionally, longitudinal changes of the volume of the cortical and sub-cortical regions as well as longitudinal change of the cortical thickness can be estimated as a single statistic (e.g. annualized atrophy rate or percent change) for each subject (dynamic features).

Both types of features were part of our previous research for image retrieval for AD [11, 12]. For completeness, we provide a short description for them in the following subsections. Considering that they bring different kind of information related to the disease, we propose to use the advantage of all of them, combine the information they carry into a single descriptor for each patient and evaluate whether they are complementary and

Step 1: Cross-sectional processing - All the available time points for all patients are indepentently processed cross-sectionally

Step 2: Template Generation - Unbiased within-subject template space and image with respect to any time point are generated for all previously independently processed time points for each subject.

Step 3: Longitudinal Pprocessing - For each time point on the basis of the template and the independent runs

Fig. 1. Main step of the processing flow using the longitudinal stream in FreeSurfer

lead to better retrieval performance. Additionally, according to our previous research [13], the SPARE-AD index provides very powerful image representation and is very successful in capturing and representing the atrophy caused by AD. We included the SPARE-AD index in this research as part of the combined descriptor.

Static Features. Static features are related to the volumetric measures of the brain regions and cortical thickness of the cerebral cortex regions, calculated at different time points [11]. They represent the current state of the patient's brain at a given moment of time, in fact, when the scan is acquired, so they are called static features. On the basis of this kind of features, we have identified the following scenarios:

1. A representation composed of volumetric measures, cortical thickness measures or a combination of volumes and cortical thickness extracted from the scan at each time point TP_Y, $Y \in \{1, 2, ..., N\}$ separately, where Y is the total number of the available time points for a patient, and in this research is 4. In this respect, a total of Y scenarios is possible.
2. A representation containing a combination of the representations extracted from the scans of the available time points (a combination of the representations obtained according to the strategy descripted in 1).

Considering this, we defined 15 scenarios with static features:

1. L-TP_1 (volume, cortical thickness and a combination of them at time point TP_1)
2. L-TP_2 (volume, cortical thickness and a combination of them at time point TP_2)
3. L-TP_3 (volume, cortical thickness and a combination of them at time point TP_3)
4. L-TP_4 (volume, cortical thickness and a combination of them at time point TP_4)

5. L-TP_1 + TP_2 (concatenated descriptors of TP_1 and TP_2),
6. L-TP_1 + TP_3 (concatenated descriptors of TP_1 and TP_3),
7. L-TP_1 + TP_4 (concatenated descriptors of TP_1 and TP_4),
8. L-TP_2 + TP_3 (concatenated descriptors of TP_2 and TP_3),
9. L-TP_2 + TP_4 (concatenated descriptors of TP_2 and TP_4),
10. L-TP_3 + TP_4 (concatenated descriptors of TP_3 and TP_4)
11. L-TP_1 + TP_2 + TP_3 (concatenated descriptors of TP_1, TP_2 and TP_3),
12. L-TP_1 + TP_2 + TP_4 (concatenated descriptors of TP_1, TP_2 and TP_4),
13. L-TP_1 + TP_3 + TP_4 (concatenated descriptors of TP_1, TP_3 and TP_4),
14. L-TP_2 + TP_3 + TP_4 (concatenated descriptors of TP_2, TP_3 and TP_4),
15. L-TP_1 + TP_2 + TP_3 + TP_4 (concatenated descriptors of TP_1, TP_2, TP_3 and TP_4).

Regarding the volumetric measures, there are 123 measures in total, while in terms of cortical thickness, we used 70 features in total. Ten of these scenarios were evaluated and the results reported in [11]. For completeness, here we provide results of the evaluation of all 15 scenarios (Table 1). Moreover, according to [37], the quality control (QC) of the processed data might influence the performance of the retrieval process, thus it is recommended to be applied. After the automated processing of the images, we detected failures in in at least one time point for some of the cases. In order to ensure a complete automatic processing and taking into account that in our research we did not have the possibility to involve a medical expert, we considered only the cases without global or regional failures in all time points (153 patients in total, 41 AD and 112 NL). In [37], the QC is not applied. Hence, for the evaluation of the scenarios for this research, we also included this step. Moreover, we applied Correlation-based Feature Selection (CFS) algorithm [38] to be able to select the most relevant features and, in the same time, to reduce the dimension of the descriptor.

Considering the separate descriptors in Table 1, volume measurements at all time points generally led to better retrieval precision. Regarding the combined descriptors, concatenation of the descriptors of the volume and cortical thickness resulted in improved results in comparison with the separate descriptors. We obtained the best results with a concatenation of volume and cortical thickness measures in the third and fourth time points. The MAP value in this case is 0.864. These results are very reasonable from clinical perspective. In fact, Anatomical changes in the relevant brain structures become more prominent in AD patients as the disease progresses. Hence, the static features extracted from scans acquired at later time points are more powerful.

Considering the dimension of the feature vector, when the information from the specific time point is used separately, the descriptor usually has between 16 and 21 features of the volume measures, from 6 to 13 for the descriptors containing cortical thickness measures, 22–25 characteristics in the concatenated vector in most cases after the application of the feature selection. Considering the combined vectors, the descriptor dimension is reduced to 27–30 features, when the information from the volumes and cortical thickness measures from two time points are used, 31–33 considering the information from three time points, and usually 33 characteristics when combining descriptors from all time points. As a result, it should be noted that a significant reduction in descriptor length is actually achieved.

Table 1. Evaluation of static features obtained by longitudinal processing.

Scenario	MAP		
	Volumes	Cortical thickness	Volumes + Cortical thickness
L-TP_1	0.798	0.771	0.811
L-TP_2	0.819	0.801	0.830
L-TP_1 + TP_2	0.818	0.799	0.823
L-TP_3	0.830	0.812	0.836
L-TP_1 + TP_3	0.834	0.807	0.834
L-TP_2 + TP_2	0.834	0.806	0.837
L-TP_1 + TP_2 + TP_3	0.830	0.806	0.838
L-TP_4	0.854	0.826	0.862
L-TP_1 + TP_4	0.849	0.824	0.863
L-TP_2 + TP_4	**0.855**	**0.827**	0.861
L-TP_3 + TP_4	**0.855**	0.826	**0.864**
L-TP_1 + TP_2 + TP_4	0.846	**0.827**	0.861
L-TP_1 + TP_3 + TP_4	0.854	0.824	0.863
L-TP_2 + TP_3 + TP_4	0.853	0.823	0.862
L-TP_1 + TP_2 + TP_3 + TP_4	0.849	0.822	0.862

Dynamic Features. Unlike static features, the dynamic features are aimed to reflect the disease progression, in fact, the speed and severity of degeneration. The dynamic features are calculated as a single statistic reduced from the temporal data for a particular measure (volume or cortical thickness of each region) taking into account all the available scans for each patient. Based on the dynamic features, the research in [12] evaluates descriptors comprised of the following statistics: rate of change (RC), percent change with respect to the value obtained from the linear fit at baseline (PC1/fit), and symmetrized percent change (SPC). Moreover, the statistics were derived from the templates generated on the bases of different number and differently spaced time points [12], as follows:

- T_123 – template based on baseline scan and the 6- and 12-month follow-ups
- T_134 – template based on baseline scan and the 12- and 24-month follow-ups
- T_234 – template based on 6-, 12- and 24-month follow-ups
- T_1234 – template based on baseline scan and the 6-, 12 and 24-month follow-ups.

According to the results reported in [12], the best value of MAP was 0.84. This value was gained in two cases, based on T1234_VolumesPCfit (descriptor comprised of PCfit changes of the volumetric measures estimated on the bases of the template generated when considering all the available scans) and T1234_VolumesSPC (descriptor comprised of SPC changes of the volumetric measures estimated on the bases of the template generated from all four time points). In this evaluation, the QC phase was also included. Regarding the feature vector dimension, in most of the cases, 19–21 features were selected.

SPARE-AD Index. The SPARE-AD index [39–41] tends to reflect the extent to which atrophy occurred in specific brain regions. It is calculated and publicly available in ADNI database.

According to our previous work and the evaluation performed in [13], the best retrieval results from the evaluated scenarios were achieved when the descriptor was composed only of the SPARE-AD. The MAP in this case was 0.81.

Combination of Static, Dynamic Features and SPARE-AD Index. Static and dynamic features carry different kinds of information extracted from the brain images. While static features represent the patient's condition at a given moment, the dynamic reflect the progress of the disease. However, none of these aspects should be overlooked. In order to determine whether they complement each other, we propose to evaluate combination of the most powerful and efficient descriptors from the scenarios based on the static and dynamic features separately. Therefore, we concatenated the appropriate feature vectors from those scenarios that led to the best retrieval results in the experiments that are based only on the static or dynamic features separately. Additionally, taking into consideration the powerful information that the SPARE-AD index carries, we also evaluated scenarios in which this index is included in the descriptor. Thus, we propose 4 scenarios based on the combination of static and dynamic features as well as SPARE-AD index:

1. SD1-Vol34 + CT34 + VolPC - concatenated measures of volume and cortical thickness of the third and fourth time point (best static features) and volume percent change (best dynamic features);
2. SD2-Vol34 + CT34 + VolSPC - concatenated measures of volume and cortical thickness from the third and fourth time point ((best static features)) and volume symmetrized percent change (best dynamic features);
3. SD3-Vol34 + CT34 + VolPC + SPARE-AD - concatenated measures of volume and cortical thickness from the third and fourth time point, the volume percent change and the SPARE-AD index;
4. SD4-Vol34 + CT34 + VolsSPC + SPARE-AD - concatenated measures of volume and cortical thickness from the third and fourth time point, volume symmetrized percent change and the SPARE-AD index;

3 Experimental Results and Discussion

Results of the retrieval based on the combined descriptors of static, dynamic measures and the SPARE-AD index are given in Table 2. This table contains the MAP value, whereas the detailed results for R-precision and precision at the fixed level are presented in Table 3. According to the results obtained from the evaluation, it can be concluded that the combination led to improvement in all cases. The best precision was obtained in the case of a combination of static measures of volume and cortical thickness from the last two time points (third and fourth), percent change or symmetrized percent change of the volume and the SPARE-AD index. The MAP in this case is 0.88 (scenario 3 and 4). Hence, the proposed scenarios outperform the scenarios based solely on static, dynamic features and SPARE-AD index. Due to the diversity of the datasets or the inclusion criteria of the subjects, we cannot perform an objective comparison with other research [14–18] focused on brain MRI retrieval for AD, although the retrieval performance improvement based on our strategy is evident.

Table 2. Evaluation of the combined descriptors.

Descriptor	MAP
SD1-Vol34 + CT34 + VolPC	0.87
SD2-Vol34 + CT34 + VolSPC	0.87
SD3-Vol34 + CT34 + VolPC + SPARE-AD	0.88
SD4-Vol34 + CT34 + VolsSPC + SPARE-AD	0.88

Table 3. Evaluation of the combined descriptors – detailed results.

Descriptor	RP	P1	P5	P10	P15	P20	P25	P30
SD1-Vol34 + CT34 + VolPC	0.83	0.90	0.91	0.90	0.89	0.89	0.88	0.87
SD2-Vol34 + CT34 + VolSPC	0.83	0.90	0.90	0.90	0.89	0.88	0.88	0.87
SD3-Vol34 + CT34 + VolPC + SPARE-AD	0.83	0.90	0.91	0.91	0.90	0.90	0.89	0.88
SD4-Vol34 + CT34 + VolsSPC + SPARE-AD	0.83	0.91	0.91	0.90	0.90	0.89	0.89	0.88

Additionally, we calculated the frequency of the selection of the features in the best two scenarios SD3 and SD4. The graphical representation of the features selected more than 50% for the retrieval based on the descriptor SD3-Vol34 + CT34 + VolPC + SPARE-AD is given in Fig. 2. Regarding the case of descriptor SD4-Vol34 + CT34 + VolsSPC + SPARE-AD, the most frequently selected measures are shown in Fig. 3. It is interesting to note that besides the known AD markers, the SPARE-AD is selected in all cases in both scenarios. In addition, most of the static and dynamic characteristics with the highest frequency of selection are from the left hemisphere. The dimension of the descriptors in most of the cases was 31–33 in both scenarios.

We believe that the improvement in longitudinal image retrieval obtained by the combination of the static and dynamic features and the SPARE-AD index occurs as a result of the different types of information these features carry. Namely, the SPARE-AD index efficiently and accurately describes the structural change in the patient's brain caused by the disease obtained on the basis of the first scan acquired for the patient. In fact, this index reflects the early stage of the disease. Then, the combination of static volumetric and cortical thickness measurements of the brain structures (which we showed to be complementary) extracted from the scans at the third and fourth time points describe the changes in the brain at fixed time points (static state) when AD is in an advanced stage. Finally, the complete progression of the disease and the degree to which it has developed (dynamic aspect) is described by the percentage change (or symmetrical percent change) of the volume of the brain structures that have been proven to be superior dynamic indicators during the retrieval.

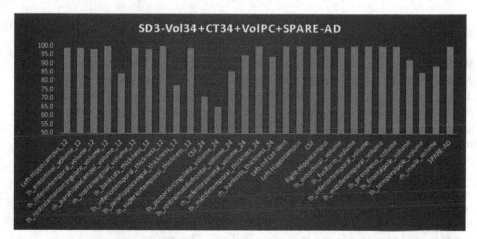

Fig. 2. The most frequently selected features (more than 50% of the cases) during the retrieval on the bases of the descriptor SD3-Vol34 + CT34 + VolPC + SPARE-AD.

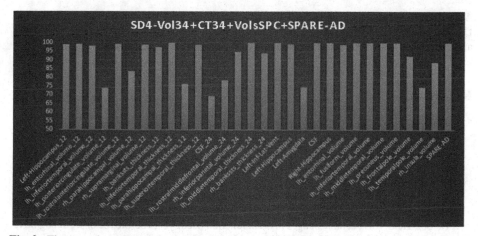

Fig. 3. The most frequently selected features (more than 50% of the cases) during the retrieval on the bases of the descriptor SD3-Vol34+CT34+VolSPC+SPARE-AD.

Combining all these aspects into one descriptor makes it comprehensive and powerful in describing the medical case based on MR images and thus leads to the best retrieval results. Such a descriptor actually allows the retrieval to be reduced to retrieve cases that have the most similar pattern of atrophy of structures affected by the disease in its early stage, the most similar anatomical condition of the brain during the development of the disease (at later time points) and the most similar dynamics of disease progression (change in brain structure/disease severity) with the case being given as a query. This is very beneficial from the medical perspective because the retrieved top similar cases bring rich clinical information to the physicians regarding the patient's current conditions, disease progression and prognosis, and/or response to the therapy.

4 Conclusion

We focused our research on improvement of the representation of longitudinal MRI for Alzheimer's disease that will properly reflect the current state and the severity of the disease and will provide semantically relevant retrieval results. For that purpose, we proposed and evaluate four scenarios based on the combination of the most powerful static and dynamic features from the previous research together with the SPARE-AD index. In fact, with the combination of the volume of cortical and subcortical regions and cortical thickness of the regions in the cerebral cortex at the later time points together with the volume percent change or symmetrized percent change, along with SPARE-AD score we superior, outperforming the scenarios based solely on static, dynamic features or the SPARE-AD index. In this case, the value of the MAP was 0.88. The number of features in the descriptor in this case was 31–33 features in most of the cases which is significantly lower than in the case of the traditional approach.

Our future work is directed towards dealing with converters, i.e., patients who converted the diagnosis during the period of examination.

Acknowledgement. Data collection and sharing for this project was funded by the Alzheimer's Disease Neuroimaging Initiative (ADNI) (National Institutes of Health Grant U01 AG024904) and DOD ADNI (Department of Defense award number W81XWH-12-2-0012).

References

1. Alzheimer's Association, 2022. 2022 Alzheimer's disease facts and figures. https://www.alz.org/media/documents/alzheimers-facts-and-figures.pdf. Accessed 30 June 2022
2. Porsteinsson, A.P., Isaacson, R.S., Knox, S., Sabbagh, M.N., Rubino, I.: Diagnosis of early Alzheimer's disease: clinical practice in 2021. J. Prev. Alzheimer's Dis. **8**(3), 371–386 (2021). https://doi.org/10.14283/jpad.2021.23
3. Winblad, B., et al.: Defeating Alzheimer's disease and other dementias: a priority for European science and society. The Lancet Neurol. **15**(5), 455–532 (2016)
4. Beason-Held, L.L., et al.: Changes in brain function occur years before the onset of cognitive impairment. J. Neurosci. **33**(46), 18008–18014 (2013)
5. Sperling, R.A., et al.: Toward defining the preclinical stages of Alzheimer's disease: recommendations from the National Institute on Aging-Alzheimer's Association workgroups on diagnostic guidelines for Alzheimer's disease. Alzheimer's & Dement. **7**(3), 280–292 (2011)

6. Wilson, R.S., Leurgans, S.E., Boyle, P.A., Bennett, D.A.: Cognitive decline in prodromal Alzheimer disease and mild cognitive impairment. Arch. Neurol. **68**(3), 351–356 (2011)
7. Agarwal, M., Mostafa, J.: Content-based image retrieval for Alzheimer's disease detection. In: 2011 9th International Workshop on Content-Based Multimedia Indexing (CBMI), pp. 13–18. IEEE (2011)
8. Cai, W., Liu, S., Wen, L., Eberl, S., Fulham, M.J., Feng, D.: 3D neurological image retrieval with localized pathology-centric CMRGlc patterns. In: 2010 17th IEEE International Conference on Image Processing (ICIP), pp. 3201–3204. IEEE (2010)
9. Mizotin, M., Benois-Pineau, J., Allard, M., Catheline, G.: Feature-based brain MRI retrieval for Alzheimer disease diagnosis. In: 2012 19th IEEE International Conference on Image Processing (ICIP), pp. 1241–1244. IEEE (2012)
10. Liu, X., Chen, K., Wu, T., Weidman, D., Lure, F., Li, J.: Use of multimodality imaging and artificial intelligence for diagnosis and prognosis of early stages of Alzheimer's disease. Transl. Res. **194**, 56–67 (2018)
11. Trojacanec, K., Kitanovski, I., Dimitrovski, I., Loshkovska, S.: Medical image retrieval for Alzheimer's disease using data from multiple time points. In: Loshkovska, S., Koceski, S. (eds.) International Conference on ICT Innovations, vol. 399, pp. 215–224. Springer, Cham (2015). https://doi.org/10.1007/978-3-319-25733-4_22
12. Trojachanec, K., Kitanovski, I., Dimitrovski, I., Loshkovska, S.: Longitudinal brain MRI retrieval for Alzheimer's disease using different temporal information. IEEE Access **6**, 9703–9712 (2017)
13. Trojacanec, K., Kalajdziski, S., Kitanovski, I., Dimitrovski, I., Loshkovska, S. and Alzheimer's Disease Neuroimaging Initiative. Image retrieval for Alzheimer's disease based on brain atrophy pattern. In: Trajanov, D., Bakeva, V. (eds.) International Conference on ICT Innovations, vol. 778, pp. 165–175. Springer, Cham (2017). https://doi.org/10.1007/978-3-319-67597-8_16
14. Leon, R.A.M., Puentes, J., González, F.A., Hoyos, M.H.: Empirical evaluation of general-purpose image features for pathology-oriented image retrieval of Alzheimer disease cases. In: CARS 2016: 30th International Congress on Computer Assisted Radiology and Surgery (2016). Int. J. Comput. Assist. Radiol. Surg. **11**, S39–S40
15. Chethan, K., Bhandarkar, R.: Hybrid feature extraction technique on brain MRI images for content-based image retrieval of Alzheimer's disease. In: Kalya, S., Kulkarni, M., Shivaprakasha, K.S. (eds.) Advances in Communication, Signal Processing, VLSI, and Embedded Systems. LNEE, vol. 614, pp. 127–141. Springer, Singapore (2020). https://doi.org/10.1007/978-981-15-0626-0_11
16. Vinutha, N., Sandeep, S., Kulkarni, A.N., Shenoy, P.D., Venugopal, K.R.: A texture based image retrieval for different stages of Alzheimer's disease. In: 2019 IEEE 5th International Conference for Convergence in Technology (I2CT), pp. 1–5. IEEE (2019)
17. Sagayam, K.M., Bruntha, P.M., Sridevi, M., Sam, M.R., Kose, U., Deperlioglu, O.: A cognitive perception on content-based image retrieval using an advanced soft computing paradigm. In: Advanced Machine Vision Paradigms for Medical Image Analysis, pp. 189–211. Academic Press (2021)
18. Kruthika, K.R., Maheshappa, H.D. and Alzheimer's Disease Neuroimaging Initiative: Multistage classifier-based approach for Alzheimer's disease prediction and retrieval. Inform. Med. Unlocked **14**, 34–42 (2019)
19. Kruthika, K.R., Maheshappa, H.D. and Alzheimer's Disease Neuroimaging Initiative: CBIR system using capsule networks and 3D CNN for Alzheimer's disease diagnosis. Inform. Med. Unlocked **14**, 59–68 (2019)
20. Veitch, D.P., et al.: Understanding disease progression and improving Alzheimer's disease clinical trials: recent highlights from the Alzheimer's disease neuroimaging initiative. Alzheimers Dement. **15**(1), 106–152 (2019)

21. Imtiaz, B., Tolppanen, A.M., Kivipelto, M., Soininen, H.: Future directions in Alzheimer's disease from risk factors to prevention. Biochem. Pharmacol. **88**(4), 661–670 (2014)
22. Chatterjee, P., et al.: Diagnostic and prognostic plasma biomarkers for preclinical Alzheimer's disease. Alzheimers Dement. **18**(6), 1141–1154 (2022)
23. Simrén, J., et al.: The diagnostic and prognostic capabilities of plasma biomarkers in Alzheimer's disease. Alzheimers Dement. **17**(7), 1145–1156 (2021)
24. Nanni, L., et al.: Comparison of transfer learning and conventional machine learning applied to structural brain MRI for the early diagnosis and prognosis of Alzheimer's disease. Front. Neurol. **11**, 576194 (2020)
25. Wang, Y., et al.: Diagnosis and prognosis of Alzheimer's disease using brain morphometry and white matter connectomes. NeuroImage: Clin. **23**, 101859 (2019)
26. Salvatore, C., Castiglioni, I.: A wrapped multi-label classifier for the automatic diagnosis and prognosis of Alzheimer's disease. J. Neurosci. Methods **302**, 58–65 (2018)
27. Soininen, H., et al.: 36-month LipiDiDiet multinutrient clinical trial in prodromal Alzheimer's disease. Alzheimer's & Dement. **17**(1), 29–40 (2021)
28. Meyer, P.F., et al.: INTREPAD: a randomized trial of naproxen to slow progress of presymptomatic Alzheimer disease. Neurology **92**(18), e2070–e2080 (2019)
29. Rebsamen, M., Suter, Y., Wiest, R., Reyes, M., Rummel, C.: Brain morphometry estimation: from hours to seconds using deep learning. Front. Neurol. **11**, 244 (2020)
30. Alzheimer's Disease Neuroimaging Initiative: ADNI (2017). https://adni.loni.usc.edu/. Accessed 30 June 2022
31. FreeSurfer. https://surfer.nmr.mgh.harvard.edu/. Accessed 17 June 2022
32. Han, X., et al.: Reliability of MRI-derived measurements of human cerebral cortical thickness: the effects of field strength, scanner upgrade and manufacturer. Neuroimage **32**(1), 180–194 (2006)
33. Reuter, M., Schmansky, N.J., Rosas, H.D., Fischl, B.: Within-subject template estimation for unbiased longitudinal image analysis. Neuroimage **61**(4), 1402–1418 (2012)
34. Zhao, W., et al.: Automated brain MRI volumetry differentiates early stages of Alzheimer's disease from normal aging. J. Geriatr. Psychiatry Neurol. **32**(6), 354–364 (2019)
35. Schwarz, C.G., et al.: A large-scale comparison of cortical thickness and volume methods for measuring Alzheimer's disease severity. NeuroImage: Clin. **11**, 802–812 (2016)
36. Voevodskaya, O., Simmons, A., Nordenskjöld, R., Kullberg, J., Ahlström, H., Lind, L., Wahlund, L.O., Larsson, E.M., Westman, E. and Alzheimer's Disease Neuroimaging Initiative: The effects of intracranial volume adjustment approaches on multiple regional MRI volumes in healthy aging and Alzheimer's disease. Front. Aging Neurosci. **6**, 264 (2014)
37. Trojacanec, K., Kitanovski, I., Dimitrovski, I., Loshkovska, S.: The influence of quality control on the image retrieval: application to longitudinal images for Alzheimer's disease. In: Proceedings of the 14th International Conference for Informatics and Information Technology, pp. 37–42 (2017)
38. Hall, M.A., Holmes, G.: Benchmarking attribute selection techniques for discrete class data mining. IEEE Trans. Knowl. Data Eng. **15**(6), 1437–1447 (2003)
39. Toledo, J.B., et al.: Relationship between plasma analytes and SPARE-AD defined brain atrophy patterns in ADNI. PLoS ONE **8**(2), e55531 (2013)
40. Davatzikos, C., Xu, F., An, Y., Fan, Y., Resnick, S.M.: Longitudinal progression of Alzheimer's-like patterns of atrophy in normal older adults: the SPARE-AD index. Brain **132**(8), 2026–2035 (2009)
41. Habes, M., et al.: White matter hyperintensities and imaging patterns of brain ageing in the general population. Brain **139**(4), 1164–1179 (2016)

Architecture for Collecting and Analysing Data from Sensor Devices

Dona Jankova[1]($\boxtimes$), Ivona Andova[1], Merxhan Bajrami[1], Martin Vrangalovski[1], Bojan Ilijoski[2], Petre Lameski[2], and Katarina Trojachanec Dineva[2]

[1] LOKA, 350 2nd Street, Suite 8, Los Altos, CA 94022, USA
{dona,ivona,merdzhan,martin}@loka.com
[2] Ss. Cyril and Methodius University, Skopje, North Macedonia
{bojan.ilijoski,petre.lameski,katarina.trojacanec}@finki.ukim.mk

Abstract. A sensor device measures one or many health indicators for the user wearing it. The vital parameters are monitored and displayed to an individual or to a medical person in real or orchestrated time. This paper focuses on creating a general architecture that enables collecting, analyzing and transmitting data depending on the user needs. It can be used to track changes in vital parameters, without the need of the individual being physically present. With the focus on smartwatch data, as an example of a wearable device, this paper shows the possibilities for its application in finding accuracy of true authentic emotions that a person with autism spectrum disorder is feeling, by measuring the person's vital parameters.

Keywords: Wearable devices · Sensors · Application · Data · Infrastructure · Framework

1 Introduction

In the past few years, a rapid development of wearable technology devices and their application in many areas, especially the healthcare industry, was witnessed [28]. With the intense development of information and communication technologies, the wearable technology ignited a new way of human-computer interaction [11,20]. A wearable device is defined as a smart electronic device, small enough to be worn as an accessory, embedded in clothing or implanted in the user's body, with the ability to measure, collect, display and even transmit health data in real time, through sensor integration [21,32]. One example is for medical purposes, where the physician would use this data to track the consumer's vital signs [4,13] such as: temperature, blood pressure, blood oxygen, heart rate, physical movement and electrodermal activity of the heart, muscles, brain and skin. This paper proposes an application architecture that can work for any sensor measuring biomedical parameters, regardless of whether it is wearable or not, like in the case of [2] for analyzing volatile organic compounds (VOCs) in the exhaled breath. The architecture can ingest the sensor data, preprocess

K. Zdravkova and L. Basnarkov (Eds.): ICT Innovations 2022, CCIS 1740, pp. 121–132, 2022.
https://doi.org/10.1007/978-3-031-22792-9_10

them and output some results to the user. In the diagram shown below (see Fig. 1), the candidate represents the person from whom the data are being collected. A sensor device can be: a camera that inspects and observes the person's movements, an audiometer that measures the sound decibels, smart glasses that give information about vision and motions etc. All of these can help people with diseases such as in the case of Parkison's [31].

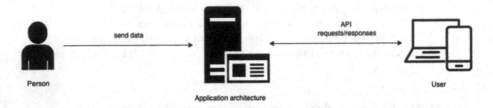

Fig. 1. High level approach on architecture diagram (Data collected from sensors worn by a person, is sent to our application architecture to process it and send the results represented in a website to the user)

After the application starts ingesting real time data, it can preprocess it, store or send the results to the client regarding the application implemented logic or user needs (see Fig. 1 - API requests/responses). Another thing that this architecture can implement is alerting someone when a certain threshold is reached or some requirements are met [18].

2   Related Work

One way of reshaping the future is using robots in everyday life. Paper [30] proposes a socially adaptable framework for human-robot interaction examining the interaction between a human and a robot in the role of a small child. Each session consists of two parts: one is when the robot can move, and the other part is when the robot can only sit and move its arms and head. The human goal is to animate the robot and examine how often the robot reaches the limit of animation saturation. In paper [1], a novel mathematical model based on multiple robot therapy for children with ASD is proposed. The model employs the robot as a therapist, without distractions from the environment, to improve the autistic child's skills. Two methods of robot interaction were conducted: with and without interaction between the robot and the child. The idea of this paper was to develop a single mathematical model for adaptive multi-robot based therapy. In this way the need for constant human therapist supervision is reduced.

Another way is to improve the utilization of wearable device data using resources and services from Amazon Web Services (AWS) by creating an architecture as proposed in article [10]. It can be used to help people with autism and their caregivers as described in paper [25], by monitoring data from multiple signals and informing the caregiver of their present feeling.

Looking from a different perspective, creating a mobile application can be very useful and also a fun way to help people with autism. It is described in the papers [14–17]. Paper [17] focuses on creating a mobile charades-style game, Guess What?. This game challenges the child to act out instructions displayed on the smartphone held to the care provider's forehead, while the parent tries to guess the emotion the child is expressing (for ex. surprised, scared, or disgusted). During this time, the app collects and processes a video recording of the child to detect emotion. This indicates what the child enjoys and feels more comfortable doing. The integration of emotion classifiers into the system is required to provide real-time feedback and adapt game difficulty. The existing emotion recognition platforms do not perform well for children with ASD, which is why the authors of the paper [16] presented a new way of extracting frames and labeling emotions in those frames in order for them to be used for training new and improved emotion classifiers. Labeling those images is better described in paper [15] using several algorithms for extracting semi-labeled frames from these videos. The common emotion classifiers are tested and described in the paper [14] using the small dataset gathered from this game. The findings suggest that the classifiers that already exist are suited for the general neurotypical children and cannot be used for digital autism treatment approaches. When more data will be generated from videos, paper [15] findings will be used to create a new dataset that will be used for training new classifiers suited for everyone.

When it comes to dealing with emotions, the paper [22] has a goal of teaching the robot to understand the emotions and help the children with ASD when dealing with a difficult situation. First, they created a representation of a child's behavioral state using Lattice Computing (LC) models together with machine learning techniques. Then they used video data from the children during robot interaction sessions to analyse those behavioral states. The conclusion of this paper is that animations and usage of LEDs tends to attract the children's attention making them lean their heads toward the robot more often. This kinds of researches can help psychologists to improve the results of the robot-assisted sessions.

For physically demanding industries sensor integrated devices can be used for monitoring biomedical changes of workers' behaviour and health, such as in the case of construction workers [8, 26].

What all of these studies can use is a general architecture for processing the data and showing the results to the user which eliminates the time spent on implementation and they can focus more on improving their model. That is the main reason why this paper proposes the implementation.

3   Architecture

3.1   Collecting Data

The architecture flow begins with collecting data. For that purpose a sensor device connected to the internet is needed in order to send the data to some kind of end point or store them in a database. One device can have multiple sensors [5,

9, 27]. With the aim of utilizing the architecture for monitoring a medical disorder such a ASD, a wearable device was put on a participant's wrist with 4 sensors: accelerometer (measures changes in speed of movements), photoplethysmograph (measures blood volume pulse - BVP), electrodermal activity of the skin and temperature sensor. Additionally from the BVP sensor another measure was extracted, that is the heart rate. All this data is sampled from the sensors at different frequencies. Every 2 s, all the data collected during that interval is sent to an end point.

3.2   Data Preprocessing

All the data that is collected is some kind of time series data, which means that in order to use it, they need to be preprocessed [19]. First, if there are any faulty values, they are removed. Then there is a need for matching the sampled frequencies of the different sensors in the same device. Each sensor is sampled at a different interval, therefore the data are needed to be duplicated in order to match the highest sample frequency. After that was sorted out, a statistical analysis was needed in order to create features that fit the previously created model's input. In this case, every attribute represents some kind of statistical analysis of the sensor data during that 2-second interval. It includes statistics such as min, max, average, median absolute deviation etc.

3.3   Analyzing Data and ML Models

After creating the features, they are fed into a model that was pre-trained on the same kind of sensor data. This can work on a wide range of different models such as: classification, clustering, regression etc., depending on the specific user needs, in order to derive a conclusion. Here, a classification approach was implemented, having an emotion as an output (see Fig. 2) [29].

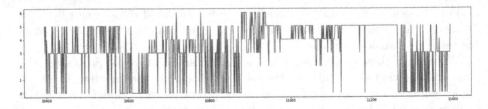

Fig. 2. Representation of how often the model predicts changes in emotion during time

Based on the performances, model stacking approach was chosen as the best one: ModelStacking(L0: Knn/SVM/LDA/RandomFores/XGBoost/ dTree/AdaBoost, L1: Xgboost) with accuracy 0.82, precision 0.78 and recall 0.71.

On Fig. 2 we can see the output emotion (Y-axis) through time (X-axis) predicted by the model. This visualization can help understand the data better and the changes in emotion (every number from 0 to 6 represents a different emotion on the Y-axis) through time predicted by the model.

3.4   Visualization

The results can be stored in another database on the cloud. Besides that, they can be streamed to a client host, depending on the needs. It can be a Web application or a mobile application that displays the results. Also, with the usage of Lambda functions, processed data can also be displayed in order to show valuable information to the user. For example Fig. 3 shows average sensor data and results from a neural network model that predicts emotions from the sensor data through time averaged on 2 s interval. Every sensor is shown with different color, X, Y and Z are metrics from the same sensor showing the direction of movement.

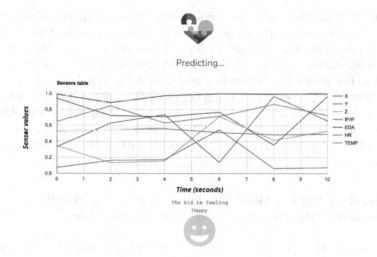

Fig. 3. Model output per sensor measurements averaged over a 2 s interval

3.5   Application Flow

Application Architecture Implemented on Amazon Web Services (AWS). This paper shows an architecture implemented on Amazon Web Services. Although, the same can be done in any other cloud computing platforms such as: Google Cloud Platform and Microsoft Azure. This implementation shows how any wearable device can be used as a source of ingesting data.

The architecture diagram on Fig. 4 shows a high-level approach of implementing the infrastructure for the wearable devices. There is a wearable devices stack

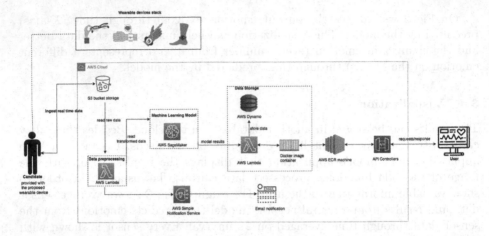

Fig. 4. Flow diagram of the implemented architecture on AWS

that consists of multiple wearable devices, where each of them is similar to the others and can be easily replaced depending on the user's needs. The application user has to establish a connection with the sensor device. Once the application user (parent, doctor, trainer, supervisor etc.) has successfully established the connection, the sensor will be sending real time data. The data will be handled by AWS services: S3 bucket for storing the data, Lambda for data preprocessing, Sagemaker for creating machine learning models used for prediction. After the model outputs the results, the application user has the option to choose how to handle the data: store it in AWS DynamoDB database or display it on the monitor.

AWS Services Explanation. The following section shows all the resources that were used for the whole process in the architecture flow:

Amazon Simple Storage Service (Amazon S3) [12] is a scalable object storage service where the raw data files from the sensors are put.

AWS Lambda in this case has multiple purposes. It is used as a compute service that preprocesses the data, in order to transform them for model input. It also reads the model output and can store it into the database or stream the data depending on user needs. Another function Lambda does is triggering the docker image building process.

Amazon DynamoDB is a NoSql database that delivers single-digit millisecond performance at any scale and is used to avoid processing of duplicate files. This database is chosen because it provides high throughput at a very low latency which is very important for reading data if the application user chooses to display the data.

Amazon Simple Notification Service (Amazon SNS) is a fully managed messaging service for both application-to-application (A2A) and application-to-person (A2P) communication that is used to send alerts to the application user for the connection status of the wearable device. It will inform whether the connection is established or not.

AWS Cloud Control API is a set of common application programming interfaces (APIs) that make it easy for developers and partners to manage the lifecycle of AWS and third-party services. It is responsible for handling API calls, such as requests and providing the application with responses.

AWS SageMaker is a cloudformation tool for machine learning purposes, used for building and training machine learning models. Depending on the project purpose multiple models can be trained on the Sagemaker instance or the instance itself can be used for statistical data analysis that can be provided to the application user.

A Docker [7,24] *image* is a file used to execute code in a Docker container. Docker images act as a set of instructions to build a Docker container, like a template. It contains application code, libraries, tools, dependencies and other files needed to make an application run.

AWS EC2 machines [3,12,23] provide scalable computing capacity in the Amazon Web Services (AWS) Cloud. An EC2 instance in the architecture will be used for building and running the docker image.

Architecture of the Web Application Used for Children's with Autism.
One way of using this architecture is in a web application developed for transmitting health data in real time from children with autism spectrum disorder to medical professionals, enabling health tracking without the individual being physically present [6]. People with autism are obstructed to understand, process and communicate emotions to other people, which explains the importance of emotional vital signs monitoring, in order to assess the individual's well-being. The data is being collected through a smartwatch containing sensors as an example of a wearable device throughout a therapy session with a doctor. Using a machine learning trained model the application can predict how the candidate is feeling during the session. The application is built and deployed on AWS and is used for visualizing the data and showing results of the model to medical personnel regarding the candidate's feelings (see Fig. 5).

The RESTful application is built with the Flask framework. Once the doctor runs the application, one will have the option to establish a device connection. Since every sensor will be generating raw data at a different frequency, data pre-processing is needed in order to match each sensor to the one with highest frequency. Every 2 s the data collected during that interval are being sent and handled from an API gateway in real time, where via POST routes the application will analyze the data. For that a script will generate statistics (features)

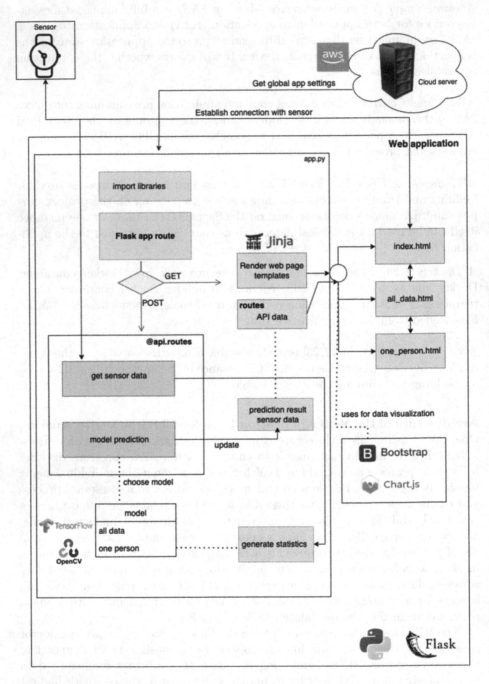

Fig. 5. Web architecture diagram that covers the whole process, from sensor connection to data visualization and the technologies that are used.

creating the data format that fits the model input. Next, the doctor can choose a model approach that will be used for predicting emotions. Once the doctor chooses it, the application will render a page where the preprocessed data and the model output will be visualized. Google Chart.js is used for data visualization and Bootstrap framework for styling. For rendering the templates via GET routes of the API, Jinja is used. The front pages that the doctor will control will be visualizing the data in real time and displaying the output model prediction.

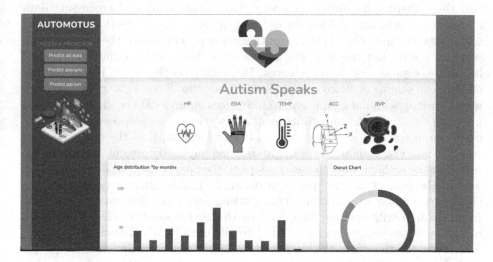

Fig. 6. Application user interface for the homepage, where sensor connection is established.

4   Conclusion

What does it take to form the new tomorrow and how do these novelties blend into what is known today? The answer is through changes. Change is a necessity. Change is a vital process, in order to build a better tomorrow. Because the increased sensor installations into objects in the human surroundings experience an exponential growth and sensors are starting to get embedded in all kinds of devices, it is essential to make a greater use out of their data. Hence, this paper focuses on providing a platform for collecting data from sensor devices (both wearable and not), making them meaningful. Data can be interpreted differently, depending on the application's context, therefore the architecture is not limited to the domain usability or appliance. The concept is the same: architecture that collects and transmits health data from sensor devices in real time and displays it to a person of interest (the user using the device or another persona - depending on the specific logic/need), so an interpretation can be made. The proposed architecture covers all the challenges related to sensor utilization, and its flexible and adjustable concept can solve any challenges with little or no adaptations. For example, If the user has a machine learning model and wants to view patterns in customer's behavior, they can make their model a central part in the architecture, and apply it to practice in order to address their operational needs and interpret their data collection. Or if a service is not appropriate for meeting the operational goals, it can be easily swapped. Simple as that, with the benefits of cloud computing. This described architecture solves the latency issues associated with data processing that can occur when sending data in real time, shaping the future in a way that transmitting and processing data in real time, without a delay, in order to make a data collection significant, is the new normal. The cloud computing concept will reduce the IT operating costs in a way that the user will get its work done, without the need of new hardware, expenses or inconveniences. Depending on the needs, it will provide effective up scaling or down scaling on the resources. Since security is a concern when working with data, Amazon Web Services ensures data confidentiality and prevents data leakage, through its available services. The proposed architecture finds use in tracking users regardless of their location, thus the gathered data can give insights, without the individuals being physically present. In view of the fact that the architecture can easily adapt to the area of interest, one adaptation was in the field of autism spectrum disorder. The architecture addressed the challenges related to gathering, transmitting and analyzing smartwatch data from children with autism, that lead to the conclusion that the authenticity of the emotions can be and is related to the data gathered from the sensors. Therefore, for this case, the data can be used for understanding how people truly feel, regardless of the emotion they are expressing on the outside.

Acknowledgement. The work in this paper was partially financed by the Faculty of Computer Science and Engineering, Ss. Cyril and Methodius University in Skopje.

References

1. Ali, S., et al.: An adaptive multi-robot therapy for improving joint attention and imitation of ASD children. IEEE Access **7**, 81808–81825 (2019). https://doi.org/10.1109/ACCESS.2019.2923678
2. Alkhouri, N., et al.: Isoprene in the exhaled breath is a novel biomarker for advanced fibrosis in patients with chronic liver disease: a pilot study. Clin. Transl. Gastroenterol. **6**(9), e112 (2015)
3. Bhise, V.K., Mali, A.S.: Cloud resource provisioning for Amazon EC2. In: 2013 Fourth International Conference on Computing, Communications and Networking Technologies (ICCCNT), pp. 1–7. IEEE (2013)
4. Boudargham, N., Abdo, J.B., Demerjian, J., Guyeux, C.: Exhaustive study on medical sensors. In: International Conference on Sensor Technologies and Applications (2017)
5. Brida, P., Krejcar, O., Selamat, A., Kertesz, A.: Smart sensor technologies for IoT. Sensors **21**(17), 5890 (2021)
6. Cabibihan, J.J., Javed, H., Aldosari, M., Frazier, T.W., Elbashir, H.: Sensing technologies for autism spectrum disorder screening and intervention. Sensors **17**(1), 46 (2016)

7. Cito, J., Ferme, V., Gall, H.C.: Using docker containers to improve reproducibility in software and web engineering research. In: Bozzon, A., Cudre-Maroux, P., Pautasso, C. (eds.) ICWE 2016. LNCS, vol. 9671, pp. 609–612. Springer, Cham (2016). https://doi.org/10.1007/978-3-319-38791-8_58

8. Edirisinghe, R.: Digital skin of the construction site: smart sensor technologies towards the future smart construction site. Eng. Constr. Archit. Manag. **26**(2), 184–223 (2019). https://doi.org/10.1108/ECAM-04-2017-0066

9. Elayan, H., Shubair, R.M., Kiourti, A.: Wireless sensors for medical applications: current status and future challenges. In: 2017 11th European Conference on Antennas and Propagation (EUCAP), pp. 2478–2482. IEEE (2017)

10. Engdahl, S.: Blogs (2008). https://aws.amazon.com/blogs/industries/improving-the-utilization-of-wearable-device-data-using-an-aws-data-lake/

11. Ernst, T., et al.: Sensors and related devices for IoT, medicine and s mart-living. In: 2018 IEEE Symposium on VLSI Technology, pp. 35–36. IEEE (2018)

12. Garfinkel, S.: An evaluation of Amazon's grid computing services: EC2, S3, and SQS (2007)

13. Howard, R.M., Conway, R., Harrison, A.J.: A survey of sensor devices: use in sports biomechanics. Sports Biomech. **15**(4), 450–461 (2016)

14. Kalantarian, H., et al.: The performance of emotion classifiers for children with parent-reported autism: quantitative feasibility study. JMIR Ment. Health **7**(4), e13174 (2020)

15. Kalantarian, H., et al.: Labeling images with facial emotion and the potential for pediatric healthcare. Artif. Intell. Med. **98**, 77–86 (2019)

16. Kalantarian, H., Jedoui, K., Washington, P., Wall, D.P.: A mobile game for automatic emotion-labeling of images. IEEE Trans. Games **12**(2), 213–218 (2018)

17. Kalantarian, H., Washington, P., Schwartz, J., Daniels, J., Haber, N., Wall, D.P.: Guess what?: towards understanding autism from structured video using facial affect. J. Healthc. Inform. Res. **3**, 43–66 (2019). https://doi.org/10.1007/s41666-018-0034-9

18. Kostikis, N., Rigas, G., Konitsiotis, S., Fotiadis, D.I.: Configurable offline sensor placement identification for a medical device monitoring Parkinson's disease. Sensors **21**(23), 7801 (2021)

19. Kotsiantis, S., Kanellopoulos, D., Pintelas, P., et al.: Handling imbalanced datasets: a review. GESTS Int. Trans. Comput. Sci. Eng. **30**(1), 25–36 (2006)

20. Lee, J., Kim, D., Ryoo, H.Y., Shin, B.S.: Sustainable wearables: wearable technology for enhancing the quality of human life. Sustainability **8**(5), 466 (2016). https://doi.org/10.3390/su8050466

21. Loncar-Turukalo, T., Zdravevski, E., da Silva, J.M., Chouvarda, I., Trajkovik, V., et al.: Literature on wearable technology for connected health: scoping review of research trends, advances, and barriers. J. Med. Internet Res. **21**(9), e14017 (2019)

22. Lytridis, C., et al.: Behavioral data analysis of robot-assisted autism spectrum disorder (ASD) interventions based on lattice computing techniques. Sensors **22**(2), 621 (2022)

23. Pham, T.P., Ristov, S., Fahringer, T.: Performance and behavior characterization of amazon EC2 spot instances. In: 2018 IEEE 11th International Conference on Cloud Computing (CLOUD), pp. 73–81. IEEE (2018)

24. Rad, B.B., Bhatti, H.J., Ahmadi, M.: An introduction to docker and analysis of its performance. Int. J. Comput. Sci. Netw. Secur. **17**(3), 228 (2017)

25. Ramasubramanian, K., Venkateswarlu, L., Lavanya, M.K., Unnati, L.: Emotional perception of individuals with autism spectrum disorder through machine learning and smart watch. Turk. J. Comput. Math. Educ. **12**(13), 7217–7225 (2021)

26. Rao, A.S., et al.: Real-time monitoring of construction sites: sensors, methods, and applications. Autom. Constr. **136**, 104099 (2022)
27. Iqbal, S., Mahgoub, I., Du, E., Leavitt, M.A., Asghar, W.: Advances in healthcare wearable devices. NPJ Flex. Electron. **5**(1), 1–14 (2021)
28. Shen, G.: Recent advances of flexible sensors for biomedical applications. Prog. Nat. Sci. Mater. Int. **31**(6), 872–882 (2021)
29. Siddiqui, U.A., et al.: Wearable-sensors-based platform for gesture recognition of autism spectrum disorder children using machine learning algorithms. Sensors **21**(10), 3319 (2021)
30. Tanevska, A., Rea, F., Sandini, G., Cañamero, L., Sciutti, A.: A socially adaptable framework for human-robot interaction. Front. Robot. AI **7**, 121 (2020). https://doi.org/10.3389/frobt.2020.00121
31. Vera Anaya, D., Yuce, M.R.: Stretchable triboelectric sensor for measurement of the forearm muscles movements and fingers motion for Parkinson's disease assessment and assisting technologies. Med. Devices Sens. **4**(1), e10154 (2021)
32. Wanjari, N.D., Patil, S.C.: Wearable devices. In: 2016 IEEE International Conference on Advances in Electronics, Communication and Computer Technology (ICAECCT), pp. 287–290. IEEE (2016)

Education

Adapting a Web 2.0-Based Course to a Fully Online Course and Readapting It Back for Face-to-Face Use

Katerina Zdravkova(✉) [iD]

Faculty of Computer Science and Engineering, Ss. Cyril and Methodius University in Skopje, Skopje, Macedonia
katerina.zdravkova@finki.ukim.mk

Abstract. COVID-19 pandemic has dramatically reshaped educational strategies and the methods of their delivery. It has also affected those courses that incorporate many social media online activities. This paper introduces the major strategies that were used to adapt an already stable Web 2.0-based course to a fully online course. A comparison of teaching and learning methods and outcomes before and after the transition is presented, illustrated by a series of selected decision factors that are discussed in more detail. It is additionally supplemented by students' evaluation of the course organization and its quality in parallel with the teacher's impression of the amount of acquired knowledge and the ways students received their passing grade. Expecting a soon return to in-class education, the course will have to undergo a new in-depth adaptation taking into account the experiences gained during fully online learning and the fact that students will have to re-learn how face-to-face education functioned.

Keywords: Course adaptation · Pre-COVID learning and teaching · Post-COVID learning and teaching

1 Introduction

Computer ethics course started as a philosophy of computing course back in 2003. During its long history, it evolved from a Web 1.0 supported to a Web 2.0 course that reinforces collaboration and social media based learning, teaching and knowledge acquisition [1]. Since academic 2011/12, most activities, including various forms of collaborative content creation have been performed predominantly online [2]. Applied teaching strategies included traditional lectures, discussions, written and audio-visual sources, and role-playing [3]. Traditional face-to-face interactions in the classroom were essential during lectures, and especially as a round-up of broad team projects [2].

COVID-19 pandemic crisis significantly disrupted organization of whole education. It needed an instant transition towards fully remote education in higher education and hybrid schooling for younger students [4]. For many educators, such an unexpected stress afforded "a golden opportunity to rethink what matters most in education" [4]. It also considerably affected computer ethics course. Since the last face-to-face delivery

K. Zdravkova and L. Basnarkov (Eds.): ICT Innovations 2022, CCIS 1740, pp. 135–146, 2022.
https://doi.org/10.1007/978-3-031-22792-9_11

ended in January 2020, there was a long period ahead for rethinking how to adapt it to fully online mode.

The adaptation was organized taking into account the experiences of educators worldwide, the experience of own colleagues, as well as own experience during the two courses in the summer semester of academic 2020/21. They will be introduced briefly in the second section of the paper. The third section presents the last pre-COVID version of the course in parallel with the adaptations initiated by the pandemic. The fourth section is dedicated to detailed analysis of time series emphasizing the alterations caused by course adaptation and the modifications due to students' oversaturation with other curricular and non-curricular obligations. The fifth part shows the evolution of students' opinions about the course, complemented by the teacher's impression of the ways in which students received the desired grade. Learned lessons will be crucial for readapting the course back to face-to-face use. They are the main topic of the paper conclusions.

2   Education: Disruption and Response

UNESCO estimated that around 85% of all the learners were not attending school in April 2020 [5]. Full closure lasted in average 19.20 weeks [5]. While Australia, Belarus, Burundi, Iceland, Russian Federation, Sweden and US continued with face-to-face education in primary and secondary schools, most other countries immediately shifted to online education [6]. This was a very stressful task for all. Academic institutions had to establish a convenient and stable environment for remote learning, teaching and assessment. With no time for detailed planning, the "great experiment" didn't work smoothly [7]. Teachers had to reinvent themselves and their teaching habits to overcome the sudden shock and enable normal education [8], hoping that the "rockiest days of remote learning" will soon be behind them [7]. Students had to start learning and preparing their assignments without being supported by their teachers and schoolmates [9]. Even the institutions with the most enthusiastic staff and with best students faced many practical and human-centered challenges, mainly because the transition was done without any warning and prior preparation [10]. Challenges among others included: lack of infrastructure for remote education [11, 12], little or no practical experience to deal with technological challenges [13], low awareness of technological challenges [14], no motivation to study remotely [15], parents' frustration and disability to help [16], and last but not least, insufficient accessibility options of learning management systems and video teleconferencing tools that disable information access of students with communication and cognitive disabilities [17].

After selecting the most suitable synchronous online teaching platform, universities had to promptly prepare teachers for basic features of technology tools. In spite of the great effort of schools and state authorities, more than 90% of teachers indicated that these crash courses were not meaningful and sufficient [18]. By the end of the first semester, most teachers and students gained the necessary experience for technology-mediated learning [19]. One of the major reasons for the progress was the increased teacher-student interaction [19]. Many schools suggested adaptation of courses to fit to fully remote education, including revised course content, realistic expectations, stimulated creativity, increased flexibility, powered by the ahead planning and testing [20].

2.1 FCSE Transition to Online Education

Faculty of Computer Sciences and Engineering (FCSE) instantaneously integrated web conferencing system BigBlueButton (BBB, https://bigbluebutton.org/) with Moodle learning management system (https://moodle.org), which has already been intensively used. After a short training session aiming to introduce BBB features, teachers were prepared for synchronous online teaching. Student assessment was set in the dedicated browser Safe Exam Browser (SEB, https://safeexambrowser.org), which disables URL or search field, back/forward navigation and reload. Monitoring of activities during the exam was enabled by applications ManyCam (https://manycam.com/) or DroidCam (https://droidcam.en.softonic.com/), students select depending on their preferences.

It was soon noticed that students' interest for attend synchronous learning considerably decreased (see Table 1). Presented data are based on Moodle's report logs of all 674 students that enrolled software engineering course in academic 2020/21. Their main excuse were overlapping obligations due to modified lecture schedule. However, they claimed that they regularly participated in asynchronous learning. Report logs show that in March around half of all students viewed recorded lectures. By the end of the semester, number of viewed recording decreased more than three times.

Table 1. Lecture attendance and recording view during Software Engineering course in 2020.

Week starting on	16.03	23.03	30.03	06.04	13.04	20.04	03.05	10.05
Lectures viewed	410	309	270	233	184	176	203	192
Recordings viewed	359	245	210	174	152	117	143	118

All software engineering teachers had an impression that their interaction during online lectures was poorer than the interaction during face-to-face classes, in spite of all the efforts to initiate communication, including interesting discussions, remarks or announced small questions related to last sentences of the lecture. It was even most embracing when the renowned software developers were invited to share their knowledge with the students. One of the main reason for the lack of interaction was the fact that students were entering the BBB room without activating the microphone, so the only medium for communication was the BBB chat area.

Apart from five-factor protection against student cheating: identification of students, virtual inspection of the room where they take the exam, SEB, listening to students during the whole exam, and careful observing of their activities, teachers' impression after the first semester was that the success rate increased in most courses, probably as a result of a communication via alternative communication media. Even placing a mirror behind the students' back didn't help much.

Lack of social interaction during online classes [21] and the risk of various forms of cheating, including contract cheating [22] were the major objectives to adapt computer ethics course in the academic 2020/21. Experience with other courses caused additional adaptations. They are the topic of the next section of this paper.

3 Evolution of Computer Ethics Course at FCSE

Computer ethics course started back in 2003, as part of the pedagogy focus groups courses recommended by IEEE and ACM joint task force on computing curricula [23]. It covered most of the topics of social and professional issues area, which were at that time part of computer science body of knowledge [23].

In the first years, course was enrolled by few dozens of students. They were regularly attending the lectures and actively participating to all discussions. Each academic year, two experts from academia and/or practice were invited to share their knowledge and skills. Students appreciated these invited talks, prepared questions in advance, making the event pleasant for themselves, the invited speaker and the teacher. Since 2005, more sophisticated video beams enabled presentation of dedicated films, TEDx talks and short videos that additionally enrich the course. Interactivity was the key component of all lectures. The assessment was also fully interactive. In the beginning, students were supposed to individually prepare and present two essays selected from a list of proposed topics. Lectures and essays were available from a Web 1.0 course site. In 2005, former Institute of informatics started using Moodle, enabling additional interactive and user-generated features. Moodle became the official learning management system of FCSE in 2011. Since 2012, student assessments have consisted of the following activities:

1. A role-playing game, which is organized in global discussion forums where students defend their role and private forums where students of one team communicate and deliver their individual and team reports [2];
2. Collaborative project, which was in the beginning done using Moodle's wiki module. Due to various incidents like fictitious re-editing and accidental or intentional obstructions of already created content it was replaced by team projects realized within private discussion forums;
3. Individual assignment on around 10 topics. Distribution of topics depended on the remainder of dividing student ID with an integer that is defined according to number of enrolled students;
4. Journal created throughout the whole course in which students collect the news related to course syllabus and concisely explain the selected ones;
5. Individual oral discussions related to role-playing game and team project;
6. Active participation during in-class lectures.

User-generated content in role-playing game and collaborative project, as well as individual assignment and journal were already online, so they did not require further adaptation. On the other hand, all face-to-face activities needed a substantial change.

To motivate students attend the lectures and actively participate in the initiated discussions, teacher started using asynchronous learning by announcing forthcoming topics, invited speakers and dedicated films, TEDx topics and videos (see left side of Fig. 1). The announcements were delivered as e-mails. Invited speakers bios, film thrillers and videos were published on the course site (see right side of Fig. 1).

To reduce the potential negative impact of already observed decreased interaction during software engineering course to final grade, weighting factors of all face-to-face activities were reduced. This modification had a positive effect on student success rate and average grade.

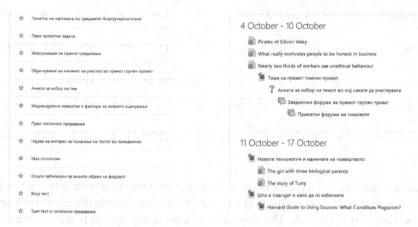

Fig. 1. First announcements and important external links.

4 What has Changed Due to Online Teaching and Learning?

As announced in the previous section, first observed difference was related with students' success rate, average results (see Table 2). Since 2012, number of students who had no activity has increased from only 7.00% in academic 2012/13 to disappointing 15.18% in 2016/17. The situation improved when the course was switched to online mode. The probable reason is that the students had to stay at home, their social activities were reduced to minimum and social media activities within the course were a direct continuation of their everyday communication.

More intensive activity resulted in better results. Although the success rate of active students oscillated between 70.63% in 2013/14 and 86.00% in 2012/13, the average result has steadily deteriorated reaching modest 63.86 points out of 100 in 2017/18. It was absolutely the best last year. Results in 2021/22 are not final. Idle students will complete their obligations during the summer break. It will increase the success rate and decrease the average result, which will again be superior to pre-COVID years.

Table 2. Time series of basic statistical data.

Basic data	2012	2013	2014	2015	2016	2017	2018	2019	2020	2021
Enrolled students	100	143	217	82	112	123	191	150	162	524

(*continued*)

Table 2. (*continued*)

Basic data	2012	2013	2014	2015	2016	2017	2018	2019	2020	2021
Completed the course	86	101	163	66	81	104	146	120	140	416
Inactive students	7	16	19	9	17	15	22	17	13	37
Success rate	86.00	70.63	75.12	80.49	72.32	84.55	76.44	80.00	86.42	79.39
Average results	72.12	68.20	69.07	72.34	72.16	68.43	63.86	72.26	78.82	77.92
Inactivity rate	7.00	11.19	8.76	10.98	15.18	12.20	11.52	11.33	8.02	7.06

Attendance rate, which was never higher that 54.73%, except in 2012, when the course was enrolled by one of the best generations of the former Institute of informatics also increased (see Table 3). Unfortunately, teacher's impression is that online presence was declarative only. Namely, since the first delivery, the key feature of computer ethics course was its intensive interaction. With all the efforts to attract students actively participate during lectures, they were either completely idle, or reacted to provocative questions with short tweet-like comments in the BBB chat.

Interaction within role-playing improved reaching in average more than 5 posts per student. It might be caused by splitting broad forums into several smaller ones with less participants [2]. Participation in private forums remained unchanged.

Interaction was acceptable during oral discussions related to role-playing and team project and particularly during invited talks. Although fewer students actively participated, all the questions, answers and conclusions were reflecting students' ability to critically think and articulate own thoughts.

Table 3. Time series of student activities and achievements.

Activities and achievements	2012	2013	2014	2015	2016	2017	2018	2019	2020	2021
Attendance rate	76.29	50.43	47.03	47.74	52.12	47.81	37.36	54.73	56.65	67.06
Discussion forum posts	312	561	687	414	413	426	673	439	764	2571
Posts per active student	3.35	4.42	3.47	5.67	4.35	3.94	3.98	3.30	5.13	5.28
Discussions in private forums	395	595	735	427	397	709	736	588	606	2103
Private posts per student	2.12	2.34	1.86	2.92	2.09	3.28	2.18	2.21	2.03	2.16

(*continued*)

Table 3. (*continued*)

Activities and achievements	2012	2013	2014	2015	2016	2017	2018	2019	2020	2021
Oral discussions	103	115	121	38	50	47	66	19	49	102
Oral discussions rate	51.50	40.21	27.88	23.17	22.32	19.11	17.28	6.33	15.12	9.73
Submitted assignments	90	98	172	67	82	104	149	118	139	455
Submission rate	96.77	77.17	86.87	91.78	86.32	96.30	88.17	88.72	93.29	93.43
Assignment grades	56.66	65.54	68.77	73.01	72.19	67.85	64.82	74.19	76.42	54.42
Submitted journals	84	91	165	67	80	95	157	122	132	443
Journal rate	90.32	71.65	83.33	91.78	84.21	87.96	92.90	91.73	88.59	90.97
Journal grades	80.12	84.69	80.38	83.88	85.81	89.22	86.97	90.94	90.00	85.97

Online course stabilized the assignment submission rate, returning it back to over 90% among the active students. The same trend was noticed with the journal submission rate. Plagiarism in the individual assignments slightly increased, while until 2020/21, ghostwriting remained almost identical reaching around 5% of all delivered essays and journals.

This academic year, ghostwriting seriously increased. Apart from a former student, whose name and surname was explicitly present in the metadata of more than 50 journals, new application for journal comparison revealed the presence of additional paper mills that produce different journals by randomly reorganizing the news extracted from the same pool of news. The application was tested for previous collections of journals, locating one to two clusters with in average of 3.47 journals with a similarity rate higher than 30%. This year, there were five such clusters, each consisting of 7 and 8 students. Similarity rate of their journals was above the threshold. Such cheating rate is almost certainly a consequence of the enormous number of enrolled students.

5   How Did Students' Opinion Evolve?

Since the first delivery of computer ethics course, student surveys have been the best catalyst for course reorganization. To check students' opinion about course organization and the contribution of various course activities, statistical and written data from the period between 2012 and 2015 were compared with the same data from 2016 and 2020, and finally this academic year. They unite the feedback of 93 students from the first period, 113 from the second, and 92 from 2021/22.

The approval rate remained stable in all three periods, with the greatest approval of 88.04% this year (see Fig. 2). Although during pre-COVID period there were minor changes, full appreciation this year is the highest, reaching the impressive 63.04%.

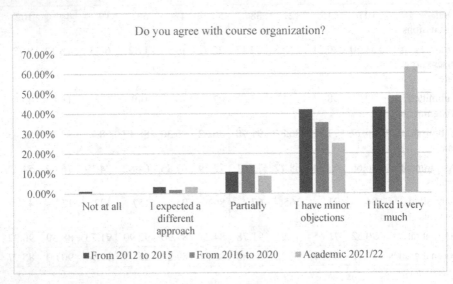

Fig. 2. Students' approval of course organization.

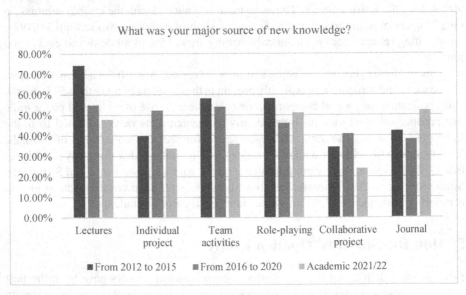

Fig. 3. Students' impression about their major source of new knowledge.

Students estimate the most important source of their new knowledge. Expectedly, their earlier impression that the lectures were the most valuable decreased (Fig. 3). This coincides with the persistently lower lecture attendance rate, which was only 11.96% in academic 2021/22, compared to 24.20% from 2012 to 2015. In the written part of the survey, students reacted that it was due to inconvenient time slots. For example, this year first lecture block was from 8am to 10am, the second from 7pm to 9pm. Teacher shares the same feelings, particularly because the earliest and the latest time slots were on Thursdays, hopefully for different groups of students. In spite of lower attendance rate, satisfaction with the lectures was constantly very high, ranging between 84.51% from 2016 to 2020 to 87.78% from 2012 to 2015.

Most interactive activities from the COVID-19 period were not appreciated as much as previously. Role-playing remained stable with 51.09% approval rate, almost equally with the debatable journal due to ghostwriting. At the same time, students showed great admiration for participating in this activity. It has increased from earlier 62.38% to very high 73.91% this academic year, which is among the best grades in the survey. The same trend was evident with the collaborative project, which was accepted in this form by 72.83% of all students. Team activities were not separately evaluated, because they are a complementary part of role-playing and collaborative project.

Individual assignments were evaluated almost equally as previously (Fig. 4). The major reason for the more or less identical grade is the same nature of the obligation, a completely online activity, which was not affected by the course adaptation. The redirection of good grades from "I made them because they are interesting" to "I made them quite easily" depended mainly on the topic assigned to each student. This year, 30 different research themes were defined. With so many themes, probability to fulfil the taste of the majority of students is much lower than with 10 themes, which are distributed using the selection strategy based on division of student's ID number.

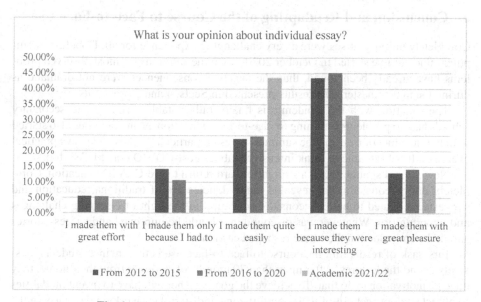

Fig. 4. Students' impression about individual assignments.

Students' written comments coincided with teacher's impression that the computer ethics course is enrolled by three types of students: excellent, regular and stubborn.

Excellent students are ambitious and compete with colleagues aiming to get the best ranking. They usually work alone trying to finish their obligations as soon as possible, much before the deadline. Role-playing game in which every post is instantly graded is their preferred activity. They obey the defined rules and avoid cheating. If they plagiarize, plagiarism mainly includes literal translation or wrong citation. After the first private warning, they excuse themselves and never repeat the mistake. They gave the best grades to all the questions in the survey.

Regular students finish their obligations timely without an ambition to get the best grade. They usually work in pair with a colleague from the same generation. They also obey the rules as much as possible. Literal translation was frequently noticed during role-playing. After being privately warned, they try to avoid it. Their evaluation of course organization and most activities was favorable.

Stubborn students start working when the deadline is approaching. They are mutually connected via social media private groups opened by students from previous generations. Their preferred activities are the individual assignments and the journal. Plagiarism in a form of literal translation is frequent. After the warning, they try to avoid it by poor paraphrasing. They are keen in hiring ghostwriters, particularly for the journal. Ghostwriting is synchronized in the private social media groups, which are self-organized student paper mills. Many individual essays and journals suspected for ghostwriting were submitted from identical IP addresses, confirming that the clusters exist. Most stubborn students simply ignored the survey. Those who answered it reacted that the course was too demanding and that they expected a different approach.

6   Conclusion and Readapting of the Course to Face-to-Face Use

Completely online courses were a very challenging experience for all. Hopefully, computer ethics course started in October 2020, when the solutions for most observed problems have already been found, thus the negative consequences were not dramatic as during the first semester. The results presented in Sects. 4 and 5 prove this.

The situation with the pandemic is finally rather stable. Primary and secondary schools had no problems returning to face-to-face education. Many faculties also started with hybrid education during the summer semester, particularly those that have practical exercises. It is high time to think intensively about post-COVID era at FCSE too.

New normal should not be a straightforward return to pre-COVID education. Completely online education has revealed many deficiencies of traditional education and suggested the need to better reconsider education in the light of emerging challenges and opportunities. Whatever was discouraging should be improved and success stories should be embedded in the future reshaped curricula.

First task of readapting the course to face-to-face use is to convince students massively attend the lectures. After more than 30 months of studying from their home, they have no motivation to do that. To achieve the goal, teacher will have to invent modalities to attract students get out of their comfort zone and start their working day early in the morning. Attractive lectures with many interesting and intriguing news and videos will probably convince better students not to miss them.

The complete online modality will be avoided as much as possible. When students are muted and their cameras are turned off, the teacher has a feeling of talking to a brick wall. Such an impression is demotivating and frustrating. However, the tutorials should remain online. During COVID era, they were massively visited. By enabling students listen to other colleagues' problems, online tutorials will contribute to resolving of many dilemmas students were simply not aware of.

Invited talks with experts, particularly those who cannot be physically present will be organized in the classroom as hybrid events. They will be open not only to students who enrolled the course, but also to wider faculty audience. Active in-class participation will be additionally awarded with bonus points. These awards will stimulate interaction and convince students attend lectures again.

In the COVID period, interaction in the computer ethics was sacrificed. It will be difficult to return it back to previous level. The best remedy is to start initiating in-class discussions about the topics that are common to Z generation. Consultations with master students and younger colleagues will be fruitful to discover which that topics are. Best consultants will be those young people who have recently successfully finished the course and still remember how it looked like. The effort in the first weeks of the new readapted course will be crucial. No matter how hard this task will be, if positive effect is achieved, it will pay off in the long run.

Acknowledgement. This work was supported in part by grants from the Faculty of Computer Science and Engineering, Ss. Cyril and Methodius University in Skopje.

References

1. Zdravkova, K.: Reinforcing social media based learning, knowledge acquisition and learning evaluation. Procedia Soc. Behav. Sci. **228**, 16–23 (2016)
2. Zdravkova, K.: Managing a successful educational role-playing game. In: Rocha, A., Adeli, H., Dzemyda, G., Moreira, F. (eds.) Information Systems and Technologies, vol. 469, pp. 433–443. Springer, Cham (2022). https://doi.org/10.1007/978-3-031-04819-7_41
3. Friedman, A., Cosby, R., Boyko, S., Hatton-Bauer, J., Turnbull, G.: Effective teaching strategies and methods of delivery for patient education: a systematic review and practice guideline recommendations. J. Cancer Educ. **26**, 12–21 (2011). https://doi.org/10.1007/s13187-010-0183-x
4. Azorín, C.: Beyond COVID-19 supernova. Is another education coming? J. Prof. Capital Commun. (2020)
5. UNESCO: Education: From disruption to recovery. https://en.unesco.org/covid19/education response. Accessed 30 June 2022
6. UNICEF: Time to roll out education's recovery package. https://en.unesco.org/news/time-roll-out-educations-recovery-package. Accessed 30 June 2022
7. Hobbs, T., Hawkins, L.: The results are in for remote learning: it didn't work. The Wall Street Journal, 5 (2020). https://www.wsj.com/articles/schools-coronavirus-remote-learning-lockdown-tech-11591375078. Accessed 30 June 2022
8. König, J., Jäger-Biela, D., Glutsch, N.: Adapting to online teaching during COVID-19 school closure: teacher education and teacher competence effects among early career teachers in Germany. Eur. J. Teach. Educ. **43**(4), 608–622 (2020)

9. Dorn, E., Hancock, B., Sarakatsannis, J., Viruleg, E.: COVID-19 and student learning in the United States: the hurt could last a lifetime. McKinsey & Company, 1 (2020)
10. Tsai, C.H., Rodriguez, G.R., Li, N., Robert, J., Serpi, A., Carroll, J.M.: Experiencing the transition to remote teaching and learning during the COVID-19 pandemic. IxD&A **46**, 70–87 (2020)
11. Misirli, O., Ergulec, F.: Emergency remote teaching during the COVID-19 pandemic: parents experiences and perspectives. Educ. Inf. Technol. **26**(6), 6699–6718 (2021). https://doi.org/10.1007/s10639-021-10520-4
12. Cenedese, M., Spirovska, I.: Online education of marginalized children in North Macedonia and Italy during the COVID-19 pandemic. Dve domovini (2021)
13. Barrot, J.S., Llenares, I.I., del Rosario, L.S.: Students' online learning challenges during the pandemic and how they cope with them: the case of the Philippines. Educ. Inf. Technol. **26**(6), 7321–7338 (2021). https://doi.org/10.1007/s10639-021-10589-x
14. Mbunge, E., Akinnuwesi, B., Fashoto, S., Metfula, A., Mashwama, P.: A critical review of emerging technologies for tackling COVID-19 pandemic. Human Behav. Emerg. Technol. **3**, 25–39 (2021)
15. Rahiem, M.: Remaining motivated despite the limitations: university students' learning propensity during the COVID-19 pandemic. Children Youth Serv. Rev. **120**, 105802 (2021)
16. Stassart, C., Wagener, A., Etienne, A.: Parents' perceived impact of the societal lockdown of COVID-19 on family well-being and on the emotional and behavioral state of Walloon Belgian children aged 4 to 13 years: an exploratory study. Psychologica Belgica **61**(1), 186 (2021)
17. Zdravkova, K, Krasniqi, V.: Inclusive higher education during the Covid-19 pandemic. In: 2021 44th International Convention on Information, Communication and Electronic Technology, pp. 833–836. IEEE (2021)
18. Marshall, D.T., Shannon, D.M., Love, S.M.: How teachers experienced the COVID-19 transition to remote instruction. Phi Delta Kappan **102**(3), 46–50 (2020)
19. Oliveira, G., Grenha Teixeira, J., Torres, A., Morais, C.: An exploratory study on the emergency remote education experience of higher education students and teachers during the COVID-19 pandemic. Br. J. Edu. Technol. **52**(4), 1357–1376 (2021)
20. Cruickshank, S.: How to adapt courses for online learning: a practical guide for faculty. HUB Johns Hopkins University (2020)
21. Adnan, M., Anwar, K.: Online learning amid the COVID-19 pandemic: students' perspectives. Online Submission **2**(1), 45–51 (2020)
22. Lancaster, T., Cotarlan, C.: Contract cheating by STEM students through a file sharing website: a Covid-19 pandemic perspective. Int. J. Educ. Integr. **17**(1), 1–16 (2021). https://doi.org/10.1007/s40979-021-00070-0
23. ACM, Computer Curricula 2001: Computer Science. https://www.acm.org/binaries/content/assets/education/curricula-recommendations/cc2001.pdf. Accessed 30 June 2022

Challenges and Opportunities for Women Studying STEM

Mexhid Ferati[1]([⊠]), Venera Demukaj[2], Arianit Kurti[1], and Christina Mörtberg[1]

[1] Linnaeus University, Växjö, Sweden
{mexhid.ferati,arianit.kurti,christina.mortberg}@lnu.se
[2] Rochester Institute of Technology – Kosovo, Prishtina, Kosovo
venera.demukaj@auk.org

Abstract. Gender stereotypes in Science, Technology, Engineering, and Math (STEM) education and careers are widely present, especially in countries with emerging economies. Making the youth interested in STEM education and careers is an important goal set by the European Commission. Consequently, understanding the obstacles youth face when choosing to study STEM is critical for policy interventions in closing the gender gap in STEM education and careers. To this end, in this paper we report on a study conducted to understand experiences of high-school and university students who study STEM. The results from two future workshops with students and a panel discussion with experts reveals three main challenges: institutional, design, and social challenges. For each challenge, we propose and discuss a respective solution: digital citizenship, universal design, and norm criticism. We conclude the paper with thoughts on the limitations of this study and directions in which this study could develop in the future.

Keywords: STEM · Education · Youth · Gender stereotypes · Workshop

1 Introduction

Despite women constituting half of the society, they make only 40% of the workforce [1] and their participation is especially low in STEM fields [2]. Specifically in Europe, they hold around 18% of ICT jobs and only 13% of them graduate with a degree in STEM [3]. Similar trends are shown in Kosovo. For instance, at the University of Prishtina, one-third of active students in STEM fields were women, whereas 44% graduated in STEM during academic year 2017–2020 [4]. These figures call for the necessity to close the gender gap in emerging economies [5]. The demand to equip new generations with skills in STEM has been recognized by the European Commission and especially much work should be done with the youth to catch the earlier periods when they start thinking about their careers in order to influence their choice [6]. Studies have shown that discussing this topic early on and especially during high school could show positive impact in increasing the interest of youth in STEM education and career [7].

To this end, in this paper we present the participant experiences we learned from two workshops organized with high-school and college students in Kosovo, which aimed

K. Zdravkova and L. Basnarkov (Eds.): ICT Innovations 2022, CCIS 1740, pp. 147–157, 2022.
https://doi.org/10.1007/978-3-031-22792-9_12

to discover the experiences of these students while studying STEM; identify the challenges faced by them; and employment prospects. In addition, workshop data were complemented with discussion panel with participants from government and university representatives and non-governmental civil institutions.

As an outcome of these activities, we identified three challenges, namely, institutional challenges, design challenges, and social challenges. To tackle these challenges, we propose digital citizenship, universal design, and norm critical measures, respectively.

The following is the structure of this paper. In the literature section we explore other studies contributing to this topic by understanding where the state-of-the-art lies and what is the gap. Further, we explain the method we used for collecting the data from our participants. In the results section, we explain in detail the three challenges identified in workshops. In the discussion section, we reflect on these challenges and how those could be addressed. Finally, we conclude the paper by highlighting its main contribution, limitations and how this work could further continue to develop.

2   Literature Review

Much research has been conducted in addressing the topic of gender gap in STEM education. The disproportional lower number of women in STEM has been experienced over the years and in different socio-economic contexts. Research conducted by Beede et al. [8] in the US context, indicated three main factors that contribute to the gender differences in STEM careers, namely; the need for female role models, the gender stereotyping, and a lack of family support. While women represent most of all graduates from tertiary education in most countries, this number gets inverted when it comes to STEM degrees [9]. Interestingly, based on the PISA data [10], even in countries with higher gender equality index such as Sweden and Norway, the number of graduates is far more tilted toward men, thus creating the phenomena that is known as "gender-paradox" [11].

A comparison of gender differences in student enrollment and completion of STEM in Massive Open Online Courses (MOOCs) conducted by Jiang et al. [12] provided some interesting findings that confirms the "gender-paradox" phenomena. According to the results provided by Jiang and colleagues, there is a higher probability for women to enroll in STEM MOOCs. Additionally, lower levels of gender gaps in STEM MOOCs enrollment and completion were found in countries that are high in gender inequality and poor economies. An interesting argument for the "gender-paradox" phenomena is given in the research conducted by Deming and Norey [13] where they suggest that "initially high economic return to applied STEM degrees declines by more than 50 percent in the first decade of working life". This especially applies for the more economically developed countries, thus potentially explaining the "gender-paradox" situation.

A recent study conducted by Tandrayen-Ragoobur and Gokulsing [14] in Mauritius indicated that factors that influence STEM career choices can be grouped in personal, environmental, and behavioral aspects. In the same study, authors suggest that the most important reasons for STEM degree choices are "self-efficacy and the student's academic performance in STEM subjects at secondary school level". Furthermore, their findings indicate that young women would decide to study STEM fields if they are supported by their family, the education institution, and teachers. This could be an indication that in less economically developed countries, softer factors such as family support, teacher encouragement and societal norms, have an important impact on women choosing STEM careers. The importance of the cultural context in clarifying gender differences in STEM has been also highlighted by research conducted by McDaniel [15].

When it comes to emerging economies, the analysis provided by OECD Skills and Work Blog [9] brings some valuable insights. The analysis done by Falco [16] suggests that some of the strategies for closing the gender gap are: removing gender bias in the curricula, increasing awareness on the consequences of choosing different fields, and that teachers should be trained and incentivized to attain greater gender equality during the learning activities. The importance of curriculum design and teachers for STEM career choices has been highlighted also by the work done by Knowles et al. [17].

The extensive research on gender differences in student career choices is based on the expectancy value model which posits that students' educational choices are linked with their expectancy for success in a particular activity and their subjective values assigned to the activities performed [18]. Findings from this research indicate that girls and women who have higher expectations for success in STEM subjects, as expressed through higher ratings of their own abilities in math and in science, tend to choose more science related courses and enroll in STEM majors [19, 20]. Women's motivations to study in specific fields are also linked to the occupation characteristics and life values as they are more likely to choose professions in which they work with people and balance their family life [21]. While most of these studies discuss motivations of students coming from the primary and secondary education, Appianing and Van Eck [22] focus on college students' motivations to stay in STEM studies. They found that women who reported higher expectations for success and assigned higher value to the STEM field, persisted in STEM studies [22].

These studies reveal the complexity of the topic and understanding the factors that motivate women to enroll and persist in STEM studies are important for policy interventions in closing the gender gap in STEM education and careers. Having this in mind, in the next section we elaborate our research efforts regarding the factors that impact the gender gap in STEM in Kosovo as an emerging economy.

3 Methodology

During late 2021, we conducted future workshops in Prishtina, Kosovo to understand the challenges of women studying STEM in this country. The future workshop approach aims to involve and elicit participant experiences and suggestions through five phases. Jungk and Müllert [23] developed the future workshop method to facilitate the process of involving citizens in public policy, which method also became relevant in design practices e.g., to Participatory Design [24]. The method includes five phases, namely: *preparation, critique, fantasy, realization,* and *follow up.* In the *preparation* phase, the materials such as post-it notes and pens are provided for participants and they are briefed about the topic, the process of the workshop and expected outcomes. In the *critique* phase, participants are encouraged to critique the current situation and write keywords on post-it notes. In the *fantasy* phase, then participants are supposed to imagine an alternative future without any constraints. This vision of the future then is evaluated with regards to economic, political, and other aspects that could make those realizable, which is the *realization* phase. Finally, in the *follow up* phase, main outcomes from the workshop are highlighted and action points are prepared [25].

For the first workshop, we recruited 18 female students from the technical high-school and the gymnasium in Prishtina. For the second workshop, we recruited 9 university female students of STEM subjects from three different universities in Kosovo. Signed consent forms were obtained from participants.

We initially briefed participants about the future workshop method, which was a novel method for them. We spent roughly 15 min in *preparation* phase, 45 min in *critique* phase, 45 min in *fantasy* phase (which was sort of combined with the *realization* phase, due to participants' lack of ability to fantasize about an imagined future), 10 min in *realization* phase, just to iterate if we missed anything, and finally, *follow up* phase for 30 min where we asked participants to provide highlights of the workshop, and brainstorm about solutions to the problems identified in the *critique* phase.

The same format was used in both workshops that were conducted in two different days; we initially met with high school students, and on the other day with university students. On the third day, we held a discussion panel with 6 representatives from universities, government institutions, and civil society associations.

The workshops with high-school and university students were mainly held using the Albanian language (participants' native language) to help them feel comfortable in expressing their ideas and also face less obstacles in engaging in the discussion with their peers and also researchers. The comments on the yellow notes were in mixed English and Albanian language. The panel discussion session with participants from the government, university, and non-governmental institutions was held in English. Most researchers were bilingual and could accommodate to participants' preference (Fig. 1).

Fig. 1. Glimpse from workshop sessions.

4 Results

The collected post-it notes produced during the workshops were consolidated with observational notes from researchers who had no "active" role in the workshops, except to observe and take notes. Those notes were in addition consolidated with notes generated from the panel discussion with stakeholder experts. With careful analysis of these notes, we noticed comments that were repeated and showed relevance for most participants. These comments were grouped depending on whether they fitted towards a common theme. In essence, these data uncovered the main challenges women studying STEM in Kosovo experience. These challenges are grouped in three themes: institutional, design, and social, which will be elaborated in the next section.

4.1 Institutional Challenges

Institutional challenges indicate a systematic failure in institutions to maintain an active role in providing the necessary support in terms of channeling and facilitating information flow. Even though there are government initiatives to support women to study STEM fields, information about these initiatives was not sufficient and timely for the targeted groups, namely high school girls. For example, workshop participants stated that there are scholarships offered specifically to increase the interest of women in STEM, however, the information about these scholarships did not reach them on time. Consequently, they end up not receiving the information about such initiatives and fail to apply. Therefore, girls highlighted that school administrators should be more updated and active in disseminating this type of information.

Also, workshop participants contended that there is lack of sufficient information available to raise awareness about the relevance of STEM education for future careers. Oftentimes, the existing information has not been encouraging for them as girls, due to the gendered representation of STEM professions in promotional materials and videos.

In the realm of the institutional challenges, the workshop participants also highlighted the lack of parents' engagement in school life (and therefore their lack of awareness on the STEM education) and the need for better cooperation between school administrators, teachers, and parents. Thus, institutions should build a better system of support and means to increase awareness.

4.2 Design Challenges

The second set of identified obstacles is related to how gender is embedded in design processes which affect women and men differently and, in a way, perpetuates gender stereotypes in technical schools. The workshop results reveal that in engineering and technical schools, which have traditionally been populated by boys, there is a lack of gender sensitive design of premises as their indoor facilities are designed primarily for boys without paying attention to girls' needs and desires. For example, girls who participated in the workshop shared their experiences that restrooms in their school lack basic privacy aspects that make it difficult for women to use them. The lack of such basic necessity undoubtedly influences women to enroll and to continue studying in such schools. The results from workshop discussion in Kosovo echo the global concerns about the ways in which the design is made as per *default male* or *one size fits men* approach [26] and in this way, it disadvantages women in classroom environments as well as in school premises, in general.

4.3 Social Challenges

Finally, social challenges, such as lack of encouragement from parents and teachers, pose a crucial obstacle that influences girls' and women's decision to study STEM. This is supported by existing gender perceptions as to who is appropriate to study and work in STEM fields, which reveal existing norms and values in the society. As indicated by the experiences of workshop participants, very often women/girls are not encouraged by their parents to study STEM because of the stigma associated with such fields and they do not want their daughter to choose a career that is associated with gender discrimination. This is an important finding given that girls, compared to boys, are more influenced by study recommendations given by their parents, siblings, and friends. The girls in the workshop claimed that they feel pressured from the family's expectations that they should behave in a way which is congruent with societal beliefs about gender roles. For example, girls are expected to study in fields that are more feminine and help-oriented (e.g., Education and Health) versus boys who are more science-and-labor-market oriented and choose to study in the technology and engineering fields. Findings from the workshop discussions confirm previous evidence from Kosovo whereby high school girls were on average more open and accepting of parents' suggestions, than boys; and in some instances, parents' suggestions were indeed decisive for their daughter' career choices [27].

Girls and women studying in STEM in Kosovo do not seem to receive robust support from their teachers either. The lack of support seems to be evident even among the teachers, who, perhaps influenced by society's general perception toward women studying STEM, tend to discourage girls and women from pursuing careers in STEM. In this context, the workshop participants shared classroom experiences whereby teachers/professors explained science concepts by referring to gender stereotypical examples (e.g., the centrifugal force and the washing machine). In some instances, women from the college group, reported as feeling unfit in the field of STEM when they would be the only woman sitting in the lab, classroom activities, or doing field work.

The attitudes experienced by workshop participants as to what professions are for girls and women vary also among the STEM disciplines. Gender stereotypical attitudes and behaviors were reported to be stronger toward girls in engineering and energy as compared to those in the ICT field. These nuanced experiences are on a par with statistics on gender representations in these fields; while the participation of women is increasing in the ICT sector, it is still much lower in the fields of electro energetics and civil engineering. This in turn influences women's persistence and retention in these men dominated fields; or, even if they graduate from these fields, they do not end up working in engineering jobs.

However, the women who participated in the workshop, contested existing norms in their choices of the STEM field. Butler [28] argues it seems humans are not able to deal with their everyday life without norms that govern their performances. But, the author also emphasizes that we do not have to assume norms are pre-given or fixed. Although existing norms are cited or reproduced, they are also exceeded or reworked.

5 Discussion

In this paper, we report on three challenges identified pertaining to gender in STEM education in Kosovo. In Fig. 2, we depict these challenges, but also present a corresponding solution that could be a way to address these challenges.

5.1 Digital Citizenship

To address institutional challenges, developing a digital citizenship eco-system could help improve the information flow between institutions (e.g., government agencies and schools), but also between educational institutions and end actors (e.g., schools and students). The notion of digital citizenship was introduced due to the development of digital technologies, and it has been defined in various ways. For instance, Ribble et al. [29, p. 7] defines digital citizenship "as the norms of appropriate and responsible behavior with regards to technology use". Alternatively, according to Mossberger et al. [30, p. 1] digital citizenship is defined as "the ability to participate in society online". Digital citizenship has been in focus in research for many years, but more critical views of the notion have emerged recently [31]. The critical voices contested the prevailing definition by arguing that a digital citizen is a "product of discursive, technological, legal, and political practices" [31, p. 508].

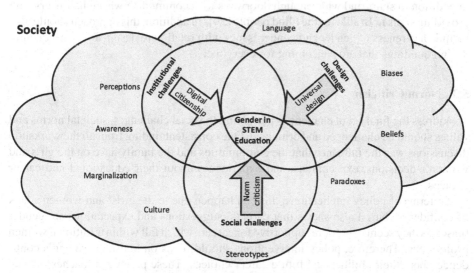

Fig. 2. Gender in STEM challenges.

Digital citizenship involves the ability to competently use and interpret the digital content. For instance, this proves challenging when the information is not provided in the language of the minorities [32] or the content could be provided in gender-biased language [33], which inadvertently could deter specific cohorts, in our case, women. Future citizens should be treated as consumer citizens and as such empower them with digital tools and content to be an active participant of the society [34]. Moreover, all involved actors in a digital society should know their roles and responsibilities to maintain a solid communication [35]. Decades long struggles to address such issues in the Swedish society have declared the feminist perspective to ensure gender equality in order to raise awareness of the existence of power structures that could be disadvantageous to women [32]. Such experiences could be relevant for traditional societies and emerging economies.

5.2   Universal Design

The design challenges could be addressed by the universal design approach to serve as a guide on how school premises are designed. Universal design consists of seven principles, which are generic enough to help design physical premises but also digital interfaces [36]. These principles can lead the process from identifying issues with existing designs to proposing solutions for physical and digital designs that are inclusive of all users regardless of their gender. Arguably, gender issues are not always included in the mainstream universal design, but the flexibility of the approach offers the possibility to complement it with Design Justice Principles to verify that design's benefits are equally shared to all involved actors [37] and that the emerging design does justice to gender [38].

A common use of universal design in school buildings is the availability of elevators in addition to stairs and wide enough doorways to accommodate wheelchair users, and providing signs in Braille for the blind people [39]. In addition, this approach should also include awareness of gender and gender issues with relation to design and consequently design buildings that are welcoming for everyone.

5.3   Norm Criticism

To address the final set of challenges pertaining to social challenges, societal norms and values should be challenged and criticized. One consistent finding from all the workshop discussions was the influence that the communities and the family have on the girls and women's decisions, expectations, and experiences about their educational and career choices.

In terms of policy implications, this can happen due to the girls' and women's lack of confidence; but, it also shows that parents' suggestions and expectations are gender biased as they seem to be more supportive for careers which fall within traditional women professions. Therefore, policy interventions should focus on boosting women's confidence about their abilities and future career choices. These policy approaches include scholarship and mentoring programs for women in STEM; role models in STEM; as well as easily accessible information about STEM importance for all the stakeholders involved, whereby developing the digital citizenship ecosystem is pivotal.

Since gender biases are reported to set in as early as pre-school age [40] addressing and mitigating these biases takes a longer period. Therefore, besides immediate, long-term impact interventions should focus on challenging and changing existing gender biases in the society by addressing the gender biases in school curriculums, textbooks, and classroom management. These interventions require a systemic approach by focusing on all age groups in the education pipeline and by including family, teachers, and the communities [27].

5.4 Reflection on the Method

As we described in the methods section, the data collection was conducted using the future workshop method. An interesting observation during the future workshop was the fact that students faced difficulty in expressing themselves in the *fantasy* phase. This phase is characterized by offering participants the opportunity to imagine a future without any constraints and limitations [41]. Participants were encouraged by researchers to fantasize about alternative views to problems identified in the *critique* phase. However, participants naturally felt the need to quickly move to the *realization* (following) phase and see only the present situation with all its constraints as a possible solution.

One explanation for this is the fact that participants reflected on existing norms and values embedded in the society and how they might govern women's choices of education. This might have limited their ability to imagine an alternative future, which consequently hampers capacity for change [42]. Another explanation was emphasized by the panel discussion with stakeholders and is related to the lack of critical thinking among students. The comment referred to the education system of Kosovo, whereby critical thinking skills of students have been underemphasized in the teaching and learning processes and students are not properly trained to think outside of the box and to project non-obvious solutions.

6 Conclusion

In this paper, we report on the uncovered experiences from participants using workshops with high school and university students in investigating challenges with studying STEM fields. The main contribution of this study is that the analysis of the workshop data identified three challenges (institutional, design, and social), for which we respectively propose three solutions (digital citizenship, universal design, and norm criticism).

The uncovered experiences from participants were rich and informative from three levels of participants: high-school students, university students, and participants from the government, university, and non-governmental institutions. Yet, we are aware of the limitation that the outcomes of the study could only reflect the condition in Kosovo and could not be generalized to a broader region, despite the presence of studies indicating similar results [20, 43, 44].

Future efforts could be directed in broadening the study with a more extended target group, namely, involving participants from elementary school as well as those participants that have graduated from STEM fields and whether they continued or not a career in STEM.

Acknowledgement. Authors would like to thank all participants who contributed with their opinions and discussions. Additionally, a special thanks goes to the Swedish Foundation for International Cooperation in Research and Higher Education (STINT) for financially supporting this research.

References

1. World Economic Forum, Global Gender Gap Report 2020. http://www3.weforum.org/docs/WEF_GGGR_2020.pdf
2. ILOStat. https://ilostat.ilo.org/techs-persistent-gender-gap/
3. Women in Digital, Scoreboard 2019, European Commission: Sweden. https://digital-strategy.ec.europa.eu/en/library/women-digital-scoreboard-2019-country-reports
4. Ministry of Education, Science and Technology & Kosovo Agency of Statistics. Education Statistics in Kosovo 2017/2018. Republic of Kosovo. https://ask.rks-gov.net/media/4146/state-arsimir-2017-2018ang.pdf
5. Central Europe's great gender opportunity. McKinsey Quarterly. https://www.mckinsey.com/featured-insights/europe/central-europes-great-gender-opportunity
6. European Commission - Horizon 2020. Science education. https://ec.europa.eu/programmes/horizon2020/en/h2020-section/science-education
7. Kim, A.Y., Sinatra, G.M., Seyranian, V.: Developing a STEM identity among young women: a social identity perspective. Rev. Educ. Res. **88**(4), 589–625 (2008)
8. Women in STEM: A gender gap to innovation. Economics and Statistics. https://files.eric.ed.gov/fulltext/ED523766.pdf
9. OECD Gender Data portal. https://www.oecd.org/employment/skills-and-work/
10. Mostafa, T.: Why don't more girls choose to pursue a science career? PISA in Focus, no.93 (2019)
11. Stoet, G., Geary, D.C.: The gender-equality paradox in science, technology, engineering, and mathematics education. Psychol. Sci. **29**(4), 581–593 (2018)
12. Jiang, S., Schenke, K., Eccles, J.S., Xu, D., Warschauer, M.: Cross-national comparison of gender differences in the enrollment in and completion of science, technology, engineering, and mathematics massive open online courses. PloS One **13**(9), e0202463 (2018)
13. Deming, D.J., Noray, K.L.: STEM careers and technological change (2018)
14. Tandrayen-Ragoobur, V., Gokulsing, D.: Gender gap in STEM education and career choices: what matters? J. Appl. Res. High. Educ. **14**(3), 1021–1040 (2022). https://doi.org/10.1108/JARHE-09-2019-0235
15. McDaniel, A.: The role of cultural contexts in explaining cross-national gender gaps in STEM expectations. Eur. Sociol. Rev. **32**(1), 122–133 (2016)
16. Gender gaps in emerging economies: the role of skills. https://oecdskillsandwork.wordpress.com/2016/07/28/gender-gaps-in-emerging-economies-the-role-of-skills/
17. Knowles, J., Kelley, T., Holland, J.: Increasing teacher awareness of STEM careers. J. STEM Educ. **19**(3) (2018). Laboratory for Innovative Technology in Engineering Education (LITEE)
18. Eccles, J.S., Wigfield, A.: Motivational beliefs, values, and goals. Annu. Rev. Psychol. **53**(1), 109–132 (2002). https://doi.org/10.1146/annurev.psych.53.100901.135153
19. Dweck, C.: Mindsets and math/science achievement. Carnegie Corporation of New York, Institute for Advanced Study, Commission on Mathematics and Science Education, New York (2008)
20. Jugovic, I.: Students' gender-related choices and achievement in physics. Center Educ. Pol. Stud. J. **7**(2), 71–95 (2017)
21. Wang, Ming-Te., Degol, J.: Motivational pathways to STEM career choices: using expectancy–value perspective to understand individual and gender differences in STEM fields. Dev. Rev. **33**(4), 304–340 (2013). https://doi.org/10.1016/j.dr.2013.08.001
22. Appianing, J., Van Eck, R.N.: Development and validation of the value-expectancy STEM assessment scale for students in higher education. Int. J. STEM Educ. **5**(1), 1–16 (2018). https://doi.org/10.1186/s40594-018-0121-8

23. Jungk, R., Müllert, N.: Future workshops: How to Create Desirable Futures. Institute for Social Inventions, London (1987)
24. Kensing, F., Munk-Madsen, A.: Participatory design: structure in the toolbox. In: PDC, pp. 47–53 (1992)
25. Vidal, R.V.V.: The future workshop: democratic problem solving. Econ. Anal. Work. Papers, **5**(4), 21 (2006)
26. Perez, C.C.: Invisible women: data bias in a world designed for men. Abrams (2019)
27. Demukaj, V., Maloku, E., Beqa, A.: Gender stereotypes and educational choices in Kosovo. Center for Social Studies and Sustainable Development LEAP (2019)
28. Butler, J.: Undoing Gender. Routledge, New York (2004)
29. Ribble, M.S., Bailey, G.D., Ross, T.W.: Addressing appropriate technology behavior. Learn. Lead. Technol. **32**(16), 6–12 (2004)
30. Mossberger, K., Tolbert, C.J., McNeal, R.S.: Digital Citizenship. MIT Press, Cambridge (2008)
31. Schou, J., Hjelholt, M.: Digital citizenship and neoliberalization: governing digital citizens in Denmark. Citizenship **22**(59), 507–522 (2018)
32. Elovaara, P., Mörtberg, C.: Design of digital democracies: performances of citizenship, gender and IT. Inf. Commun. Soc. **10**(3), 404–423 (2007)
33. von der Malsburg, T., Poppels, T., Levy, R.P.: Implicit gender bias in linguistic descriptions for expected events: the cases of the 2016 United States and 2017 United Kingdom elections. Psychol. Sci. **31**(2), 115–128 (2020)
34. SOU 2004:56, E-tjänster för alla (English: E-services for everyone). http://www.regeringen.se/content/1/c6/02/23/19/95d852b3.pdf
35. Ribble, M.: Digital citizenship in schools: Nine elements all students should know. International Society for Technology in Education (2015)
36. Story, M.F.: Principles of universal design. In: Universal Design Handbook (2001)
37. Costanza-Chock, S.: Design justice: towards an intersectional feminist framework for design theory and practice. In: Proceedings of the Design Research Society (2018)
38. Van der Velden, M., Mörtberg, C.: Between need and desire: exploring strategies for gendering design. Sci. Technol. Human Values **37**(6), 663–683 (2012)
39. Pisha, B., Coyne, P.: Smart from the start: the promise of universal design for learning. Remedial Special Educ. **22**(4), 197–203 (2001)
40. Bian, L., Leslie, Sarah-Jane., Cimpian, A.: Gender stereotypes about intellectual ability emerge early and influence children's interests. Science **355**(6323), 389–391 (2017). https://doi.org/10.1126/science.aah6524
41. Dator, J.: From future workshops to envisioning alternative futures. Futur. Res. Q. **9**(3), 108–112 (1993)
42. Hajer, M., Versteeg, W.: Imagining the post-fossil city: why is it so difficult to think of new possible worlds? Territory Polit. Gov. **7**(2), 122–134 (2019)
43. Ranković, N., Mece, E.K., Ivanović, M., Stoyanova-Doycheva, A., Savić, M., Ranković, D.: Female students' attitude towards studying informatics and expectations for future career-Balkan case. In: Proceedings of the 9th Balkan Conference on Informatics (BCI 2019), Sofia, Bulgaria, pp. 26–28. ACM (2019)
44. Putnik, Z., Štajner-Papuga, I., Ivanović, M., Budimac, Z., Zdravkova, K.: Gender related correlations of computer science students. Comput. Hum. Behav. **69**, 91–97 (2017)

Medical Informatics

Novel Methodology for Improving the Generalization Capability of Chemo-Informatics Deep Learning Models

Ljubinka Sandjakoska[1]([⊠]), Ana Madevska Bogdanova[2], and Ljupcho Pejov[3,4]

[1] Faculty of Computer Science and Engineering, UIST St. Paul the Apostle, Ohrid, Macedonia
ljubinka.gjergjeska@uist.edu.mk
[2] Faculty of Computer Science and Engineering, University SS Cyril and Methodius, Skopje, Macedonia
[3] Faculty of Natural Sciences and Mathematics, University SS Cyril and Methodius, Skopje, Macedonia
[4] Department of Chemistry, Bioscience and Environmental Engineering, Faculty of Science and Technology, University of Stavanger, Stavanger, Norway

Abstract. In the last decade, the research community has implemented various applications of deep learning concepts to solve quite advanced tasks in chemistry, ranging from computational chemistry to materials and drug design and even chemical synthesis problems at both laboratory and industrial – grades. Because of the advantages as a high-performance prediction tool in molecular simulations, deep learning is becoming far more than just a temporary trend. Instead, it is foreseen as a tool that will be essential to employ throughout tackling a range of different issues in chemical sciences in the nearest future. In this paper, we propose a novel methodology for regularization of deep neural networks used in chemo-informatics. The methodology consists of four blocks: *Class of initial conditions*; *Orthogonalization, Activation* and *Standardization*. Three graph-based architectures are developed: deep tensor neural network, directed acyclic graph and convolutional graph model. Graph-based models are more convenient for modeling molecules since the molecules and their features are often naturally represented by graphs. Several experiments are obtained on datasets from MoleculeNet aggregator: QM7, QM8, QM9, ToxCast, Tox21, ClinTox, BBBP and SIDER, for predicting geometric, energetic, electronic and thermodynamic properties on small molecules. The obtained results outperform some of the published references and give directions for further improvement. As a particular example, in one of the architectures, we have reduced mean absolute error by more than 12 times compared to conventional regression models, and more than 3 times in comparison to deep networks where the proposed methodology is not implemented.

Keywords: Deep learning · Regularization · Chemo-informatics · Molecular simulations

K. Zdravkova and L. Basnarkov (Eds.): ICT Innovations 2022, CCIS 1740, pp. 161–174, 2022.
https://doi.org/10.1007/978-3-031-22792-9_13

1 Introduction

The increased application of deep learning concepts in natural sciences in general, and chemistry in particular, can be evidenced by a simple glance over the current hot topic in the corresponding research communities [1]. A wide variety of tasks specific to conventional chemistry, such as prediction of molecular activity, toxicity [2], reactivity, protein contact; prediction of the drug-target interaction [3]; virtual screening [4]; quantitative structure activity relationship (QSAR) predictive modeling [5]; prediction of metabolism, absorption, excretion, distribution and toxicity (ADMET) properties [6], have got their substantially improved solution by using deep learning. First, the deep learning techniques are adjusted as complements to more standard (or, perhaps, more conventional) software for molecular simulations. Today, there is a tendency of developing deep architectures designed based on a detailed analysis of all factors and variables in the domain of the chemical task. Before creating deep neural networks, it is necessary to implement procedures for determining distinctive chemical features (known as featurization). Therefore, cooperation between experts in the field of machine learning and chemistry is more than necessary.

There are several reasons behind the increased need for implementation of the "deep learning" approach in chemistry. One of the most distinctive advantages is the multitask learning feature [7], which enables sharing of hidden representations in data of processing units between prediction tasks. Another argument for the justification of application is the automatic construction of complex features [8], due to the ability of deep learning to produce a hierarchical representation of a compound, with higher levels presenting more complex concepts [9].

1.1 Related Work

In reference [3] a comparative analysis between deep learning and standard machine learning techniques, such as logistic regression, kNN, binary kernel discrimination and support vector machines, is done to predict drug-target interaction. Reimplementation of some commercial software is also included to depict the advantages of deep learning. The comparison of the ROC curves indicates that deep learning architectures outperform other methods and commercial software.

Perhaps an even better example is from the field of toxicity prediction and evaluation [2], where predicting an arbitrary number of toxic effects that occur at the same time is often made possible by a deep learning architecture without having to train a separate classifier for each effect. The improvement of the performance in deployment of deep learning architectures is a result of the use of the representation learned in the multi-task environment.

Furthermore, the authors in [4] where virtual screening for compound-protein prediction is given, emphasize that the deep learning process does not require training with all the input data, because of the specifically developed ability of the network which can be trained with small mini-series. Thus, gaining computer time and memory is a certain advantage. The experiments show that deep neural networks outperform the standard CGBVS approach (Chemical Genomics-Based Virtual Screening). The cross-validation results show an accuracy of 98,2% with 4 million compound-protein interactions.

The nature of biological data, which is inherently characterized by nonlinearity, imbalance, noise, and diversity, has motivated the authors of [10] to apply deep architecture to enhance prediction. In addition, deep learning approaches, in [11], have proven to be a good tool for predicting drug-releasing microspheres in the drug formulation process. Deep learning architectures are also successful in predicting the water solubility of drug molecules. Explicitly, the authors of [12] conclude that the performance obtained by deep learning coincides with or exceeds the most modern methods according to several metric measures. This implementation reflects a specific advantage of in-depth learning such as its reliability in identifying only appropriate minimal molecular descriptors because the corresponding representations are learned automatically from the data without preprocessing the data or adjusting the hyper-parameters. The experiments in the drug discovery process have been advanced using in-depth learning to predict QSAR [5], which is another motive for applying these concepts.

From the brief overview of the application, we can conclude that deep learning is widely used in this domain. The benefits mentioned above open opportunities which will enhance in-depth research in the wide field of chemistry, that in turn is closely related to pharmaceutical research as well as to the design and application of new materials. Improvements lead to cost savings and a reduction in the number of high-cost in-vivo experiments. Researchers in the field of advanced machine learning techniques believe that the future role of deep learning will not only be as a high-performance prediction tool but also serve as a "device" for generating new hypotheses.

2 Experimental Design

2.1 Methods and Materials

2.1.1 Methods

Chemical systems are inherently complex. One should bear in mind that the science of complexity has its basic roots in fundamental chemical problems.

The range of specific chemistry-related problems that can be addressed with the proposed methodology is rather wide. For example, a wide range of molecular electronic properties can be predicted, such as molecular geometry, atomization energy, electron affinity and ionization potential, polarizability tensor components, electronic spectra and relative positions of low-lying singlet excited states, thermochemical properties, etc. As can be seen from the presented results, the methodology developed in the current study enables more accurate predictions of the above-mentioned parameters in comparison to the other methods.

The complexity of chemical systems implies complexity in their computer modeling, at most due to the variations of the impact factors on each piece of data we observe. Deep neural networks, as modeling and predictive tools, dealing with the complexity of the systems, become complex too. Increasing the complexity of the deep neural model usually means storing many hidden layers as well as excessive data parameterization. Excessive parameterization of deep neural networks, on the other hand, is often the main cause of overfitting. A network with millions of parameters can be easily overfitted. The same problem occurs in cases of insufficient data for model training. Overfitting results in a reduction of the generalization ability of the neural network. Good generalization, in the broadest sense, means that a deep neural network can deliver good performance not only on a training dataset but also on new, previously unseen data. Achieving high generalization performance is the guiding idea of this paper. Improving the generalization is usually done by regularization procedure. The regularization procedure is a dynamic process that depends on many factors. It can be derived from any supplementary technique aimed at improving the generalization of the predictive model.

In this paper, we propose a novel methodology for the regularization of deep neural networks used in chemo-informatics. We use the abbreviation ChIREG (Chemo-Informatics REGularization). The regularization is applied to different deep learning architectures. Three graph-based architectures are developed: deep tensor neural network (DTNN), directed acyclic graph (DAG) and convolutional graph model (CG). Graph-based models are more convenient for modeling molecules since molecules and their features are often naturally represented by graphs.

DTNN Architecture

The basic architecture (Fig. 1) is originally defined in reference [13]. Namely, it is a deep architecture of a neural network with tensors designed for solving several tasks. It consists of computing features on batch, data generation, embedding and gathering layer. Several parameters are included: number of features per atom/number of embeddings, number of features for each molecule after DTNN step/number of hidden neurons, a granularity of the step size of the distance matrix, a minimum distance of atom pairs (default = 1 Ångstrom), a maximum distance of atom pairs, (default = 18 Ångstrom). To build features for the central dark green atom (Fig. 1), in this deep architecture, explicit bonding information is not included. Updating the features, during the learning process in the neural network, is done with all other atoms using their corresponding physical distances. Standard procedures of training, testing and validation are comprehended.

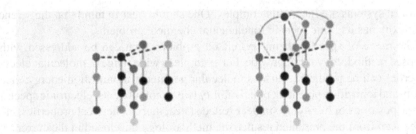

Fig. 1. DTNN architecture

DAG Architecture

These acyclic graph models describe the molecules as directional graphs, although chemical bonds have no directions. The architecture on which the ChIREG methodology is implemented is based on reference [14], where molecular properties are predicted. DAG architecture is used to build features for the central dark green atom (Fig. 2), realized with the propagation of the features through directed bonds from the farthest atom to the central atom. In all the cases, the bonds are directed towards the central atom.

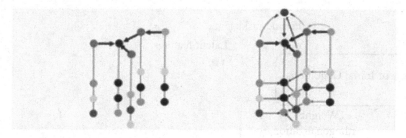

Fig. 2. DAG architecture

GC Architecture

The convolutional model treats molecules as unordered graphs. Namely, in this deep architecture, the same learning function is applied to each node (atom) and its neighbors (bound atoms) in the graph. This structure summarizes the layers of convolution in the deep grids for visual recognition. That is, it uses convolutions but at the nodes of the graph instead of the pixels in the image. The design of this model has basically been derived from reference [15] – the main difference is that when initializing the convolution layer, a predefined activation function is not used, but the ChIREG activation module is activated, which selects the most appropriate one from the search space. In the graph convolutional model features are updated by combination with neighbor atoms (Fig. 3).

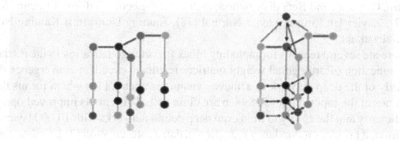

Fig. 3. GC architecture

The methodology for regularization of the previously described deep architectures consists of four blocks: *Class of initial conditions*; *Orthogonalization, Activation* and *Standardization* (Fig. 4). The basic function of the first block is to find the appropriate class of initial conditions. The class is identified by creating pseudo-tasks. The purpose of pseudo-tasks is to extract latent features of data. Using the latent features, we obtain pre-training. The pre-training directly participates in finding the appropriate class initial conditions. Knowledge of such a class will be used to obtain matrices of weights. The resulting matrices will be orthogonalized. The operations from the second block - orthogonalization are performed to obtain stability of the activations. In the third block of this methodology, there is a mechanism for the selection of the activation function. The last block is intended for standardization, which should control the activation by introducing a hyperparameter. Here we would like to emphasize that the block "class of initial conditions" is different from the initialization of the network.

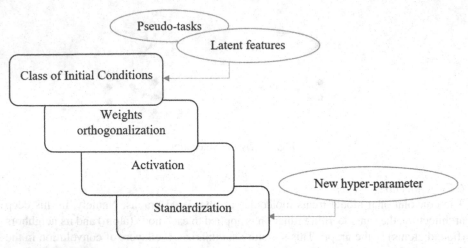

Fig. 4. Regularization methodology for improving the generalization capability

The procedure of initialization, which includes assigning values for the weights and biases for the neurons, is done by drawing values from well-known data distributions In order to find the most appropriate distribution for the biases and weights (before orthogonalization) for each data set, we obtain several experiments: initialization using Uniform, Gaussian and Beta distribution; additionally Lecun Uniform и Lecun Normal [16, 17], Xavier Uniform и Xavier Normal [18], Kaming Uniform и Kaming Normal [2] initializations.

There are several reasons for including block for orthogonalization in the methodology. Application of orthogonal weight matrices results in excellent convergence due to the ability of the deep network to achieve dynamic isometry, [19] which means that all unit values of the input-output Jacobian are close to 1. This entails improved optimization efficiency and the ability to train even deep neural networks with 10,000 layers [20]. In addition, it provides redundancy reduction in data representation [21]. Orthogonalization provides stability of the distribution of activations in different layers of the network. It is done using an algorithm known as *Orthogonalization by Newton's Iteration* (ONI) [22]. ONI is modified and the algorithm implemented in the proposed methodology does not include centering on the auxiliary parameters. It starts from the assumption that the unit values of the auxiliary parameters have a balanced distribution, to obtain the speed of convergence, but also to reduce the calculations. It is characteristic that this algorithm controls the magnitude of the orthogonality of the weights matrix by iterative "stretching" of its values to 1.

The activation block of ChIREG works with a specific mechanism for activation. First, function search space is created. The activation function $f_A(x)$, can have three general forms, the occurrence of each of which is equally probable: $f_A(x) = U_i(x)$, $f_A(x) = U_i(U_{i+j}(x))$ and $f_A(x) = B(U_i(x), U_{i+j}(x))$ where $U_i(x)$ is unary function, while $B(U_i(x), U_{i+j}(x))$ is a binary function. The indexes are integers, in the intervals $i \in (1, 15)$ and $j \in (1, 14)$. The intervals are defined by the number of unary functions that are used. The set of unary functions used in this research contains the following functions:

$0, 1, x, -x, |x|, x^{-1}, x^2, x^{\frac{1}{2}}, e^x, e^x - 1, loglog(e^x + 1), x \cdot \sigma(x)$ (known as Swish), where $\sigma(x) = (1 + e^{(-x)})^{-1}$ is sigmoid function, $x \cdot tanh(lnln(1 + e^x))$ (known as Mish) and $\frac{x}{\frac{x}{\alpha} + e^{-\frac{x}{\beta}}}, \alpha, \beta > 0$ (as Soft-Root-Sign). The set of binary function include the following functions: $U_1 + U_2, U_1 - U_2, U_1 \cdot U_2, \frac{U_1}{U_2}, U_1^{U_2}, (U_1, U_2), min(U_1, U_2)$. The activation mechanism automatically selects one of the possible activation functions from the search space, formed by the specified unary and binary functions. The search space is formed as the sum of the permutations for all three general forms in which the activation function occurs. On Fig. 5. Is given example for possible combination of the activation functions: a) $f_A(x) = U_1(x)$, b) $f_A(x) = U_1(U_2(x))$ and c) $f_A(x) = B(U_1(x), U_2(x))$.

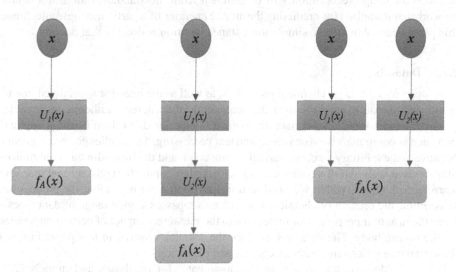

Fig. 5. Variants for combining activation functions

a) For the first general form of the activation function, defined with $f_A(x) = U_i(x)$, we have to choose one from 15 unary functions. In the language of combinatorics, permutations $P(15, 1) = \frac{15!}{(15-1)!} = 15$;

b) For the second general form, defined with $f_A(x) = U_i(U_{i+j}(x))$, where $i \neq j$, we chose 2 from 15 unary functions, or $P(15, 2) = \frac{15!}{(15-2)!} = \frac{15 \cdot 14 \cdot 13!}{13!} = 15 \cdot 14$ functions;

c) For the third general form, $f_A(x) = B(U_i(x), U_{i+j}(x))$, we use 15 unary и 7 binary functions and obtain $7 \cdot \frac{1}{15}P(15, 15) = 7 \cdot \frac{1}{15}\frac{15!}{(15-15)!} = 7 \cdot \frac{1}{15}\frac{15 \cdot 14!}{0!} = 7 \cdot 14!$ functions.

The total number of candidate functions is $15 + 15 \cdot 14 + 7 \cdot 14!$ i.e., 610 248 038 625. Due to computer resources, the number of used unary and binary functions is limited, still, the search space is large. These many possible activation functions further increase the complexity and cause difficulties in the activation mechanism. It should be emphasized that the purpose of the activation mechanism is not to find a universal, new activation function, although this is not excluded, but the most appropriate for the given task and the given data set. The activation mechanism solves a classic optimization problem, with iterative search - attempts to find the best of the candidate solutions in relation to the selected criterion.

Standardization is performed by introducing a new hyperparameter. The purpose of introducing this hyperparameter is to limit the range of values that can be received by the output. In this way, we protect the loss function from acquiring large values. It should be noted that this step of the regularization methodology is exclusively related to the problem that is being solved. For example, the standardization of a deep neural network intended for image recognition will be different from the standardization in a neural network that is applied for predicting the atomic energies of a particular molecule. Since this paper deals with atomic simulations, standardization is done in that domain.

2.1.2 Datasets

Heterogeneity of data on chemical properties, as well as the need for specialized instruments and human expertise, affect the acquisition of sufficient, authentic, accessible, consistent and non-redundant data, in contrast to other areas of deep learning application such as computer vision or speech and text processing. The challenge here is greater because of the arbitrary size, the variable connection, and the three-dimensional molecular forms as well. All this contributes to non-standard inputs for conventional machine learning techniques, which we need to transform into a form suitable for processing. In addition, the breadth of chemical research encompasses a wide range of data types - from the quantum properties of molecules to the measured impact of certain molecules on the human body. The data sets used for the research covered in this paper fall into these two categories - quantum mechanics and physiology.

We use MoleculeNet as a reference aggregator for databases and models [20]. MoleculeNet, following the example of the well-known ImageNet [21] and WordNet [22] databases, uses multiple public databases with corresponding evaluation metrics and offers high-quality open-source implementations of several previously proposed molecular marking and machine learning algorithms. The properties of over 700,000 chemical compounds are included into the datasets. The MoleculeNet datasets are incorporated into the open source DeepChem package, which is a project for which that the author himself [23] says that it aims to democratize deep learning in the natural sciences. In addition to DeepChem, MoleculeNet uses Scikit-learn and TensorFlow [24]. In the implementation of the proposed methodology in this paper we also use Keras [14], which is a deep learning API written in Python, running on top of the machine learning platform TensorFlow. The MoleculeNet datasets can be grouped according to the type of task they are used to solve. We have regression datasets and classification datasets. The golden rule in machine learning for dividing data into three subsets: for training, validation and testing is also applied in MoleculeNet in the ratio of 80/10/10. For consistency, but also relevant comparison, we use the same ratio for data sharing. The experiments included in this paper follow a different approach to data sharing. This is because the random splitting of molecular data has not been shown to be always the best way to evaluate machine learning models. In the evaluation, we use scaffold splitting, which distributes the samples based on their two-dimensional structural frames. This division tends to divide structurally different molecules into different subsets to help learning algorithms.

The QM7 and QM7b datasets contain 3D Cartesian coordinates and electronic properties of 7165 molecules, for regression. These datasets are used for predicting the electronic properties, such as Atomization energy – PBE0, Excitation energy of maximal optimal absorption – ZINDO, Highest absorption – ZINDO, HOMO – ZINDO, LUMO – ZINDO, First excitation energy – ZINDO, Ionization potential – ZINDO, Electron Affinity – ZINDO, HOMO – KS, LUMO – KS, HOMO – GW, LUMO – GW, Polarizability – PBE0, Polarizability – SCS.

QM8 is a dataset that includes low-lying singlet-singlet vertical electronic spectra of over 20 000 synthetically feasible small organic molecules. Prediction of the following features E1 – CC2, E2 – CC2, f1 – CC2, f2 – CC2, E1 – PBE0, E2 – PBE0, f1 – PBE0, f2 – PBE0, E1 – CAM, E2 – CAM, f1 – CAM, f2 – CAM are processed high-throughput across chemical space.

QM9 is used for predicting atomic properties by means of geometric properties (atomic coordinates) which are integrated into features (mu, alpha, HOMO, LUMO, gap, R2, ZPVE, U0, U, H, G, Cv).

Prediction of the barrier permeability, as a classification task, is done using the BBBP dataset which contains blood-brain barrier penetration features.

Tox21, ToxCast and ClinTox are for the prediction of toxicity.

SIDER dataset consists of drug side-effects organized into 27 system organ classes following MedDRA classifications measured for 1427 approved drugs.

2.1.3   Evaluation Measures

The proposed methodology is evaluated and compared with the benchmark models with respect to: Mean Absolute Error (MAE) when solving regression task; and the Area Under the Curve (AUC) of Receiver Characteristic Operator (ROC) (or ROC-AUC) when it comes to classification.

2.1.4   Experimental Setup

The hyperparameters are optimized by Gaussian Process Optimization in maximum of 20 iterations. The training of the models is limited and is not longer than 10 h.

3   Results and Discussion

From the obtained results we can conclude that the proposed methodology ChIREG works well in deep neural networks with tensors when solving a regression problem (Table 1) and in a convolution graph in classification (Table 2). Thus, for the data set QM7 exceeds the two reference models KRR and ANI-1. If we observe the values of the used metrics, we can see that with the application of the proposed methodology in a deep network with tensors MAE is reduced by more than 12 times compared to conventional regression models, and more than 3 times in the case of deep networks where ChIREG is not implemented. Thus, for the same QM7 data we have a drastic jump in prediction error

(Table 3. ChIREG DAG model), three times higher compared to conventional models and over 37 times higher in comparison to graph-based models. This clearly points out the fact that the ChIREG methodology is not the most suitable one for deep directional acyclic graph architecture. It is obvious that additional adjustments need to be made to make this type of model work.

Table 1. Summary of models' performance for prediction on quantum mechanics datasets - Reference vs. ChIREG models

Mean Absolute Error (MAE) in kcal/mol						
Dataset	Conventional model with best performance [14]	MAE	Graph-based model with best performance [14]	MAE	ChIREG DTNN model	ChIREG DAG model
QM7	KRR	10.220	ANI-1	2.8600	0.8096	30.5600
QM7b	KRR	1.050	DTNN	1.7700	0.0366	4.3598
QM8	Multitask	0.015	MPNN	0.0143	0.0013	0.0240
QM9	Multitask	4.350	DTNN	2.3500	2.1432	3.8600

The same trend of predictive performance improvement is followed for the other sets in this group QM7b, QM8 and QM9 (Table 2) although the differences are here smaller. It is noticeable that better performances are obtained not only for the conventional multitask network, but also for the deep MPNN (Message Passing Neural Network) architecture. Thus, for the data from the QM8 database, the use of the ChIREG methodology enables a reduction of the error from 0.0143 to 0.00129468, which is a more than 10 – fold improvement. Figure 7 depicts the difference in the performance of deep tensor neural networks without the application of the proposed methodology and with the application of the ChIREG methodology. Thus, the improvement in decreasing the predictive error is noticeable in Fig. 7, which proves the efficiency of the proposed regularization methodology.

In order to amplify and justify what has been said above in Table 2 the values of the mean absolute error (MAE) for all tasks of QM9 (prediction of atomic properties: mu, alpha, HOMO, LUMO, gap, R2, ZPVE, U0, U, H, G, Cv) are given. The results show that the deep tensor neural network with the ChIREG methodology applied exceeds the performance in 6/12 tasks. Thus, we have an improvement in the MAE value from 2.35 to 2.14. When it comes to convolution graph application, the methodology gives better results in 10/12 tasks, and the total error is reduced by 0.1731. Although the reduction is small in both cases, it is not insignificant. This indicates the regularization effect of the proposed methodology designed for solving chemo-informatics tasks.

Looking at the results given in Table 3 we can conclude that the convolution graph with the applied ChIREG methodology has significantly better performance than the standard convolution graph and the conventional reference models with the best performance for the data sets: Tox21 and SIDER. The ROC-AUC value shows an improvement

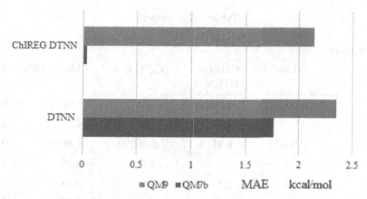

Fig. 6. Difference of performance of deep nets with tensors without application of the proposed methodology and with application of ChIREG methodology

of more than 6% or 9% for the specified sets respectively. It is noticeable that the deep networks with tensors in which the proposed methodology is applied do not give the best results in the category, although they also show improvements. For example, ChIREG DTNN model outperforms the performance of convolutional graphs: GC and ChIREG GC but does not perform better than the conventional KernelSVM on BBBP dataset. Also, the ChIREG DTNN model gives better results for SIDER data but only compared to the reference research models [14]. The performance of the Wave models for ToxCast and ClinTox is not exceeded by using the proposed methodology. This indicates the possible impact of data representation on the model architecture.

If we compare the convolutional graphs that use the proposed methodology and those that do not (Fig. 6), we can conclude that higher values for the metrics are obtained in the cases of the ChIREG model.

Table 2. Summary of models' performance for prediction on QM9 dataset for all tasks – Reference vs. ChIREG models

MAE in kcal/mol						
	Task	DTNN [14]	ChIREG DTNN	GC [14]	ChIREG GC	Min MAE
1	mu	0.244	**0.19258**	0.583	**0.425**	0.19258
2	Alpha	0.95	0.97	1.37	**1.3**	0.95
3	HOMO	0.00388	0.00278	0.00716	**0.00658**	0.00278
4	LUMO	0.00513	0.0052	0.00921	0.0093	0.00513

(continued)

Table 2. (*continued*)

MAE in kcal/mol

	Task	DTNN [14]	ChIREG DTNN	GC [14]	ChIREG GC	Min MAE
5	gap	0.0066	**0.00654**	0.0112	0.0134	0.00654
6	R2	17	**14.3**	35.9	**34.6**	14.3
7	ZPVE	0.00172	**0.0168**	0.00299	**0.00198**	0.00168
8	U0	2.43	2.5	3.41	**3.28**	2.43
9	U	2.43	2.5	3.41	**3.28**	2.43
10	H	2.43	2.5	3.41	**3.28**	2.43
11	G	2.43	2.5	3.41	**3.28**	2.43
12	Cv	0.27	**0.24**	0.65	**0.62**	0.24
	Average MAE	2.35011083	**2.14323167**	4.34779667	**4.17468833**	

Table 3. Summary of models' performance for prediction on physiology datasets - Reference vs. ChIREG models

ROC-AUC

Dataset	Conventional with the best performance [14]	ROC-AUC	Graph-based with bet performance [14]	ROC_AUC	ChIREG DTNN model	ChIREG GC model
BBBP	KernelSVM	0.729	GC	0.6900	0.7124	0.7078
Tox21	KernelSVM	0.822	GC	0.8290	0.7159	0.8919
ToxCast	Multitask	0.702	Wave	0.7420	0.7225	0.6349
SIDER	RF	0.684	GC	0.6380	0.7059	0.7378
ClinTox	Bypass	0.827	Wave	0.8320	0.8119	0.8055

4 Conclusion and Future Work

In the present study we develop a new method for regularization of deep neural networks used in chemo-informatics. The newly developed methodology consists of four blocks: *Class of initial conditions*; *Orthogonalization*, *Activation* and *Standardization*. Three graph-based architectures are developed: deep tensor neural network, directed acyclic graph and convolutional graph model. It is demonstrated that graph-based models appear to be more convenient for modeling molecules. We attribute this to the inherent natural representation of molecules and their features by graphs. The results obtained on several datasets from MoleculeNet aggregator outperform some of the published references and give directions for further improvement. In one of the developed architectures, the mean absolute error has been reduced by more than 12 times as compared to conventional regression models, and more than 3 times in comparison to deep networks where the proposed methodology is not implemented. Thus, the Chemo-Informatics Regularization (ChIREG) methodology outperforms the benchmarks in several cases. Therefore, we expect that the methodology itself and its possible further improvements can contribute to the development of more accurate predictive tools with potential application for the solution of numerous wide-range problems in the broad context of chemical sciences.

The future work can include creating a unified framework for learning latent features of the data during the determination of the initial conditions for DNN. An additional value of the work will be achieved if we examine the possibility of finding a collective previous effect to obtain explicit distinguishing of the sets of cross-referenced features.

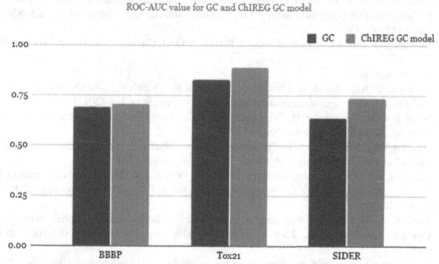

Fig. 7. Comparison of ROC-AUC value for prediction with standard convolutional graph and convolutional graph with ChIREG methodology applied

References

1. Sandjakoska, L., Bogdanova, A.M.: Deep learning: the future of chemoinformatics and drug development. In: 15th International Conference on Informatics and Information Technologies, CIIT (2018)
2. Unterthiner, T., Mayr, A., Klambauer, G., Hochreiter, S.: Toxicity prediction using deep learning. arXiv preprint arXiv:1503.01445 (2015)
3. Unterthiner, T., et al.: Deep learning for drug target prediction. Work. Represent. Learn. Methods Complex Outputs (2014)
4. Hamanaka, M., et al.: CGBVS-DNN: prediction of compound-protein interactions based on deep learning. Mol. Inf. **36**(1–2), 1600045 (2017)
5. Ma, J., Sheridan, R.P., Liaw, A., Dahl, G.E., Svetnik, V.: Deep neural nets as a method for quantitative structure–activity relationships. J. Chem. Inf. Model. **55**(2), 263–274 (2015)

6. Hughes, T.B., Miller, G.P., Swamidass, S.J.: Modeling epoxidation of drug-like molecules with a deep machine learning network. ACS Cent. Sci. 1(4), 168–180 (2015)
7. Caruana, R.: Multitask learning. Mach. Learn. 28(1), 41–75 (1997)
8. Bengio, Y., Courville, A., Vincent, P.: Representation learning: a review and new perspectives. IEEE Trans. Pattern Anal. Mach. Intell. 35(8), 1798–1828 (2013)
9. Bengio, Y.: Deep learning of representations: looking forward. In: Dediu, A.-H., Martín-Vide, C., Mitkov, R., Truthe, B. (eds.) SLSP 2013. LNCS (LNAI), vol. 7978, pp. 1–37. Springer, Heidelberg (2013). https://doi.org/10.1007/978-3-642-39593-2_1
10. Tian, K., Shao, M., Wang, Y., Guan, J., Zhou, S.: Boosting compound-protein interaction prediction by deep learning. Methods 110, 64–72 (2016)
11. Zawbaa, H.M., Szlęk, J., Grosan, C., Jachowicz, R., Mendyk, A.: Computational intelligence modeling of the macromolecules release from PLGA microspheres—Focus on feature selection. PLoS ONE 11(6), e0157610 (2016)
12. Lusci, A., Pollastri, G., Baldi, P.: Deep architectures and deep learning in chemoinformatics: the prediction of aqueous solubility for drug-like molecules. J. Chem. Inf. Model. 53(7), 1563–1575 (2013)
13. Martins, I.F., Teixeira, A.L., Pinheiro, L., Falcao, A.O.: J. Chem. Inf. Model. 52, 1686–1697 (2012)
14. https://keras.io/
15. Schütt, K.T., Arbabzadah, F., Chmiela, S., Müller, K.R., Tkatchenko, A.: Quantum-chemical insights from deep tensor neural networks. Nat. Commun. 8(1), 1–8 (2017)
16. Altae-Tran, H., Ramsundar, B., Pappu, A.S., Pande, V.: Low data drug discovery with one-shot learning. ACS Cent. Sci. 3(4), 283–293 (2017)
17. Gayvert, K.M., Madhukar, N.S., Elemento, O.: A data-driven approach to predicting successes and failures of clinical trials. Cell Chem. Biol. 23(10), 1294–1301 (2016)
18. Artemov, A.V., Putin, E., Vanhaelen, Q., Aliper, A., Ozerov, I.V., Zhavoronkov, A.: Integrated deep learned transcriptomic and structure-based predictor of clinical trials outcomes. BioRxiv, p. 095653 (2016)
19. Jain, A.N., Nicholls, A.: Recommendations for evaluation of computational methods. J. Comput. Aided Mol. Des. 22(3–4), 133–139 (2008). https://doi.org/10.1007/s10822-008-9196-5
20. Wu, Z., et al.: MoleculeNet: a benchmark for molecular machine learning. Chem. Sci. 9(2), 513–530 (2018)
21. Deng, J., Dong, W., Socher, R., Li, L.J., Li, K., Fei-Fei, L.: ImageNet: a large-scale hierarchical image database. In: 2009 IEEE Conference on Computer Vision and Pattern Recognition, pp. 248–255. IEEE (2009)
22. Miller, G.A.: WordNet: a lexical database for English. Commun. ACM 38(11), 39–41 (1995)
23. Ramsundar, B.: Molecular machine learning with DeepChem. Doctoral dissertation, Stanford University (2018)
24. Abadi, M., et al.: Tensorflow: large-scale machine learning on heterogeneous distributed systems. arXiv preprint arXiv:1603.04467 (2016)

An Exploration of Autism Spectrum Disorder Classification from Structural and Functional MRI Images

Jovan Krajevski, Ilinka Ivanoska$^{(\boxtimes)}$ ⓘ, Kire Trivodaliev ⓘ,
Slobodan Kalajdziski ⓘ, and Sonja Gievska ⓘ

Faculty of Computer Science and Engineering, Ss Cyril and Methodiuos University,
Skopje, North Macedonia
jovan.krajevski@students.finki.ukim.mk,
{ilinka.ivanoska,kire.trivodaliev,slobodan.kalajziski,
sonja.gievska}@finki.ukim.mk

Abstract. There are strong indications that structural and functional magnetic resonance imaging (MRI) may help identify biologically relevant phenotypes of neurodevelopmental disorders such as Autism spectrum disorder (ASD). Extracting patterns from MRI data is challenging due to the high dimensionality, limited cardinality and data heterogeneity. In this paper, we explore structural and resting state functional MRI (rs-fMRI) for ASD classification using available ABIDE II dataset, using several standard machine learning (ML) models and convolutional neural networks (CNNs). To overcome the high dimensionality problem, we propose a simple data transformation method based on histograms calculation for the standard ML models and a simple 3D-to-2D and 4D-to-3D data transformation method for the CNNs in ASD classification. Numerous research has been done for ASD classification using the online available ABIDE I dataset, and several with the ABIDE II dataset, the latter mostly working with single-site classification studies. Here, we take the whole ABIDE II dataset using all structural and functional raw data. Our results show that the proposed methods achive state-of-the art results of high 71.4% accuracy (functional) and 73.4% AUC (structural) compared to the best performing results in literature of 68% accuracy (functional) for ASD classification on all ABIDE dataset.

Keywords: fMRI · Autism spectrum disorder · Histogram transformation · CNN

1 Introduction

Autism Spectrum Disorder (ASD) is a heterogenous group of neurodevelopmental disorders characterized by reduced socialization and communication ability and limited and repetitive behavior, affecting nearly 1% of the world population. The manifestation of the symptoms can develop gradually with age and

---

Supported by Faculty of Computer Science and Engineering, Skopje, N. Macedonia.

K. Zdravkova and L. Basnarkov (Eds.): ICT Innovations 2022, CCIS 1740, pp. 175–189, 2022.
https://doi.org/10.1007/978-3-031-22792-9_14

varies across patients. ASD is linked to a number of genetic factors, but also to environmental factors, air pollution, toxins use in pregnancy, alcohol, drugs, pesticides, as well as various infectional and autoimmune diseases. Similarly to other neuropsychiatric conditions, ASD affects the brain information processing by altering brain connectivity. Magnetic resonance imaging (MRI) has been shown to be a useful tool for investigating such brain differences on the autism spectrum, especially using functional magnetic resonance imaging (fMRI).

The publicly available ABIDE autism neuroimaging dataset [10,11] includes two large scale collections (ABIDE I and ABIDE II) consisting of 2000+ anatomical, structural and resting-state functional MRI (rs-fMRI) images with phenotype data of ASD patiens and healthy controls from 29 different acquisition centers. The exploration of this dataset can be of great benefit for the discovery of ASD biomarkers and overcoming the challenges of reproducibility and generalizability to a larger population of neuroimaging MRI ASD studies.

The neuroimaging community has adopted and highly exploited the use of ML and deep learning methods for numerous psychiatric and neurological disorders classification tasks [3,23,36,45,47,52]. All of these methods, especially deep learning ones, try to face the numerous challenges when dealing with neuroimaging data like the high-dimensionality of the either 3D anatomical and structural data (eq. T1 weighted [12,37]), or the 4D fMRI data; or the small datasets size which rarely surpasses 100 subjects.

To overcome MRI data challenges, brain activity is usually reduced in dimension spatially or temporally by calculation of a brain connectivity correlation matrix [40] and classification models are based on those connectivity matrices [22,25,28,32]. For ASD, several studies perform classification on the transformed brain connectivity data [2,23,24,51] using atlas based regions-of-interest (ROI) functional connectivity. Furthermore, over the past few years, convolutional neural networks (CNNs) have become significantly more popular for processing MRI data [42,48]. For ASD classification, for example, [43] proposed different transformations to keep the full spatial resolution with examination of the rsfMRI temporal dimension, and train a full 3D CNN with the result of 66% accuracy on all ABIDE data. Moreover, [14] proposes a hybrid 3D CNN and 3D C-LSTM based model, and although achieves high accuracy on single-site ABIDE I collection, it reports only 58% accuracy on the multi-site ABIDE I ASD classification. [7] proposes a 4D CNNs combined with recurrent models which jointly learns from spatial and temporal data, for a result of 67% accuracy and a F1-score of 0.71. In contrast, [13] proposes a simple atlas-based 1-D CNN fMRI ASD classification with 68% accuracy. Similarly, [31] proposed a simple and efficient atlas-based hybrid 1D CNN and LSTM model for a high 79% accuracy, but tested only on single-site ABIDE I collection.

In this paper, we propose: 1) a data transformation method for the 3D structural T1 MRI and 4D rsfMRI images, with histograms calculation for an ASD classification with standard ML algorithms and 2) a 3D-to-2D data transformation for the T1 MRI and 4D-to-3D data transformation methods for the rsfMRI images, for a 2D and 3D CNNs ASD classification, correspondingly. These proposed approaches deal with the data high-dimensionality and simplify the classification directly learning imaging features, with no brain connectivity transformation or matrices calculation. We take into account the whole multi-site ABIDE II dataset to test our proposed approaches, resulting in highest 71.4% accuracy (functional) and 73.4% AUC (structural T1) for ASD classification for this dataset, which surpasses highest reported state-of-the-art accuracy of 68% (functional) of the 1D-CNN approach on the ABIDE dataset proposed in [13].

The paper is organised as follows. In Sect. 2 the dataset used in this study is presented with the data preprocessing, and data transformation with the proposed histograms calculation method, and 3D-to-2D and 4D-to-3D transformations. Afterwards, Sect. 3 details all standard ML methods and proposed CNNs for ASD classification. Results are presented in Sect. 4, reported separately for structural and functional data. Lastly, Sect. 5 draws some conclusions, and suggests some future steps forward.

2 Materials and Methods for Dataset Transformation

2.1 Dataset and Image Acquisition

The dataset used for this study is the Autism spectrum disorder (ASD) ABIDE II dataset [10], online available at http://fcon_1000.projects.nitrc.org/indi/abide/abide_II.html. The dataset consists of 19 sites ASD collections and 2 additional sites ASD longitudinal collections in total of 1114 subjects from 521 ASD patients and 593 healthy controls (age range: 5–64 years). Details for the imaging data acquisition of all collections is available at http://fcon_1000.projects.nitrc.org/indi/abide/abide_II.html. We have used the T1 structural and resting state fMRI imaging data from the 19 collections from different sites selecting 448 ASD individuals and 451 healthy controls, discarding longitudinal collections due to subjects repetition. The subjects with incomplete or missing data (e.q. subjects with only T1 structural data) were not included in the selected dataset for further processing.

2.2 Data Preprocessing

This study uses standard techniques and libraries for preprocessing T1 structural imaging and resting state fMRI data. The T1 imaging data preprocessing was done using several steps: MRtrix3 [44] mrconvert conversion from DICOM to NIfTI data type [27]; FSL [21,50] fslreorient2std reorienting to match standard FSL Montreal Neurological Institute (MNI) template to ensure same orientation for all subjects and ANTs library [5] N4BiasFieldCorrection correction for bias field/intense inhomogeneity correction; and finally FSL BET [38] skull stripping for parts analysis image removal.

FMRI preprocessing was performed with the Statistical Parametric Mapping (SPM12) [39] library in MATLAB R2018b and the CONN toolbox [49] version 18b. The T1 preprocessed data was inserted in the CONN toolbox for a fMRI preprocessing pipeline with several steps: removal of first 5 time points, realignment, unwraping; slice-timing correction; ART [4] outliers detection; images segmentation and normalization to standard MNI template; 8mm full width at half maximum (FWHM) kernel) smoothing; and 0.008–0.09 Hz frequency window band-pass filtering.

2.3 Data Transformation

The data is heterogeneous, therefore, for improving classification performance feature selection with descriptors extraction is done when using standard ML classification models. For that purpose, we propose a histograms calculation method, with standard data normalization technique that facilitates a classification algorithm training process to achieve better results. Additionally, to deal with high data dimensionality, we propose a 2D transformation image output from the 3D T1 structural imaging data as an input to a 2D CNN; as well as, a 3D transformation of the 4D rs-fMRI images for a 3D CNN.

2.3.1 Data Scaling

To deal with data magnetic resonance imaging heterogeneity, in addition to differences in maximum intensities in different images not to force erroneous classification results, it is necessary to scale each pixel/voxel in relation to other pixels/voxels in a given image. The scaling is done with

$$X_{i,c} = \frac{X_{i,c} - \min X_i}{\max X_i - \min X_i} \tag{1}$$

for each pixel/voxel intensity c ($X_{i,c}$ as mean value of its channels, here with only one channel) of image X_i.

2.3.2 Histograms Calculation

One histogram calculation of the whole image, loses pixel/voxel values information, therefore, a different way of calculating histograms is proposed. Firstly, we divide the image into 10 equal parts along each of its axes, so that each pixel/voxel belongs to exactly one subimage. Thus, from the 3D T1 images we get 1000 non-overlapping subimages, and from the 4D fMRI images we get 10000 non-overlapping subimages. From each of the subsets, we calculate a 20 bins histograms, resulting in 20 bins 1000 histograms from the 3D T1 images, and 20 bins 10000 histograms from the 4D fMRI images. In this way, the histogram values completely loose the location information of the pixels/voxels that were used to obtain the histogram, but the histogram itself retains its location relative to the other histograms of that image.

Magnetic resonance imaging (MRI) images do not have the same resolution and their resolutions differ drastically if obtained from different machines (as in our study ABIDE II dataset from 19 different collections sites), so for those higher resolution images, the values of the bins in the histograms will be on average higher than the values of the bins in the histograms for the lower resolution recordings. In general, $\sum_i hist_{small,subimg,i} < \sum_i hist_{large,subimg,i}$, where $\sum_i hist_{small,subimg,i}$ is a histogram bin of subimage *subimg* from a lower resolution image, whereas $\sum_i hist_{large,subimg,i}$ from a higher resolution image. To overcome this not to affect classification results, for each bin $hist_{img,subimg,i}$ of a histogram with 20 bins extracted from subimage $subimg$ of image img, we transform $hist_{img,subimg,i} = \frac{hist_{img,subimg,i}}{\sum_j hist_{img,subimg,j}}$ where $\sum_i hist_{img,subimg,i} = 1$. Finally, we concatenate the histograms for each image, giving a descriptor with length of 20,000 for each 3D T1 image and a descriptor with length of 200,000 for each 4D fMRI image.

2.3.3 Data Normalization

For the neural networks (NN) classification models described further in this paper and obtaining better classification results, we performed additional independent variables NN normalization. If the independent variables are not normalized and the rank (difference between the maximum and minimum value) of the two independent variables is significantly different, the weights value in the neural network will increase the influence of the independent variables by a higher order of the predicted result, which can slow down the NN training process and force a larger number of classification errors. The normalization is done for each voxel with coordinates $j = (x, y, z)$ in 3D T1 images and $j = (x, y, z, t)$ in 4D fMRI images in image i with $X_{i,j} = \frac{X_{i,j} - \overline{X_J}}{\sigma X_j}$, where $\overline{X_J}$ is the mean value of all voxels with coordinates j and σX_j is the standard deviation of all voxels with coordinates j. In this way, all voxels with coordinates j will have a mean value equal to zero and a standard deviation and dispersion equal to one.

2.3.4 3D and 4D Imaging Data Presentation in 2D and 3D Form

We use 2D CNNs further in the classification process, therefore, an adequate 2D data presentation for the 3D T1 image and 4D fMRI images is needed in 2D form. The problem is solved in the same way for both 3D and 4D images. Taking into account only 3D images, then we can see each image as a series of 2D images along the z-axis, therefore the 3D images with resolution (H, W, L) (H - height, W - width, L - length) be presented as 2D images with resolution $(H, W * L)$. Thus, each voxel with coordinates (h, w, l) bijectively is mapped into a pixel with coordinates $(h, w * L + l)$.

The same method can be applied to 4D images, looking at each 4D image with resolution (H, W, L, T) ((H - height, W - width, L - length, T - time) as a series of 3D images along the time axis. Then, we can present 4D images as 3D images with resolution $(H, W, L * T)$ which can be represented in 2D with resolution $(H, W * L * T)$. Thus, each 4D voxel with coordinates (h, w, l, t) bijectively is mapped to a 3D voxel with coordinates $(h, w, l * T + t)$ bijectively mapped again to a pixel with coordinates $(h, w * L + l * T + t)$.

Additionally, given that the fourth axis of 4D images is temporal, it is a sequence of 3D images, therefore the 4D image with resolution (H, W, L, T) can be represented into T 3D images with resolution (H, W, L) and used to train a 3D CNN.

3   Classification Models

Several standard ML classification models have been developed using the Scikit-learn library [30] in Python for this study listed in Sect. 3.1. They use the previously proposed data transformation with histogram calculation method for image representation. Additionally, using the Keras [9] and Tensorflow [1] libraries for Python, we implemented a 2D CNN and 3D CNN, which use the previously proposed 3D-to-2D and 4D-to-3D images transformation.

3.1   Standard Machine Learning Classification Models

The following standard ML algorithms performed the classification: **Gauss's naive Bayes (GB)** [34]; **Multilayer Perceptron (MP)** [33] (we choose the regularization hyperparameters L1 or L2); **K nearest neighbors (KNN)** [6] (we choose the number of K nearest neighbors as hyperparameters between 2, 5 or 7; the voting weight of each of the neighbors between uniform weight or weight proportional to the distance of the neighbor; and between one of the two algorithms available in Scikit-learn to find the closest neighbors); **Logistic regression (LR)** [19] (we choose the regularization L1 or L2); **Ridge** [18]; **Decision trees (DT)** [35] (we choose the division selection method); **Random forest (RF)** [29] (we choose: the classifiers number that use the decision tree as part of the ensemble (2, 5, 10, 50, 100 or 200); two methods for selecting a division; the minimum number of data samples that a node needs to have to stop splitting; and the maximum number of descriptors for choosing which way to split the node: 25%, 50%, 75% or 100% of the descriptors); **Extra trees (ExT)** [17] (we choose the same hyperparameters as of RF classifier); **Gradient boosting (GB)** [16] (we choose the error function the same as the LR error function or the same as the error function in AdaBoost algorithm; and the number of boosting stages is considered: 10, 50, 100, 200 or 1000); **Stochastic gradient descent (SGD)** [8] (we choose one of five different error functions to be optimized and the type of regularization: L1 or L2); and **Linear Support vector machines (Lin SVM)** [46] (We use the linearn SVM implementation in Scikit-learn using the liblinear library [15]; we consider the penalty regularization parameter: 1, 1.5 and 2).

3.2   2D and 3D Convolutional Neural Networks Classification Models

Here, we consider two types of CNNs for the ASD classification: 2D CNNs and 3D CNNs with the dataset already transformed as in Sect. 2.3 with standard convolutional, maxpooling and fully connected layers, ELU and softmax activation functions with Adam error function optimizer [26].

To overcome the problem of the relatively low number of MRI images to be used in a CNN for a deep learning classification problem and avoid overfitting, a careful NN regularization needs to be performed. First, we use regularization by reducing the dataset size in each iteration of the training epochs. Furthermore, we use dropout regularization [41] which reduces interdependence between neurons. Lastly, to speed up the training NN process, better NN generalization, and avoiding overfitting we use regularization by dataset normalization [20].

3.2.1   2D Convolutional Neural Network Architecture

We only look at the architecture of the 2D CNN that has achieved the best results. The 2D CNN has two convolutional layers with filters with dimensions $(3, 3, 1)$, two MaxPooling layers, one layer which is responsible for representing the output with resolution (H, W, C) (H - height, W - width, C - number of channels) into a vector with dimensions $H * W * C$, two fully connected layers and one output layer. Convolutional layers and fully connected layers are regularized by dataset normalization, and the output layer with resolution (H, W, C) into a vector with dimensions $H * W * C$ is dropout regularized. The complete architecture of the 2D CNN is shown in Fig. 1.

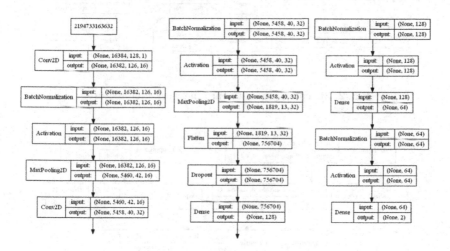

Fig. 1. 2D convolutional neural network architecture

3.2.2   3D Convolutional Neural Network Architecture

We used two 3D CNN architectures that differ in the way they treat 4D images, with the second 3D CNN being used only for 4D images. The first 3D CNN trained has two convolutional layers with filters with dimensions $(3, 3, 3, 1)$, two MaxPooling layers, one layer which is responsible for representing the output with resolution (H, W, C) into a vector with dimensions $H * W * C$, four fully connected layers and one output layer. Convolutional layers and fully connected layers are regularized by dataset normalization and the output layer with resolution (H, W, C, T) (H - height, W - width, C - channels, T - time) into a vector with dimensions $H * W * C * T$ is dropout regularized. The complete architecture of the first 3D CNN is shown in Fig. 2.

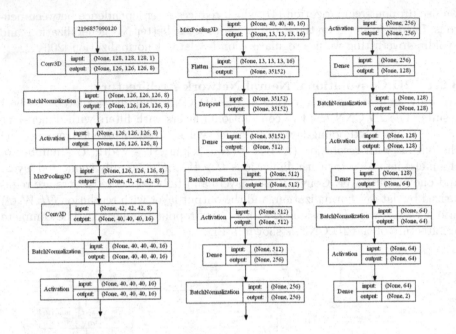

Fig. 2. 3D convolutional neural network architecture

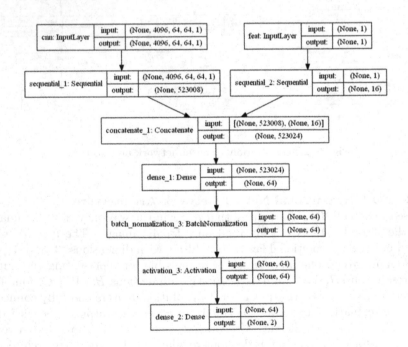

Fig. 3. 3D non-sequential convolutional neural network architecture

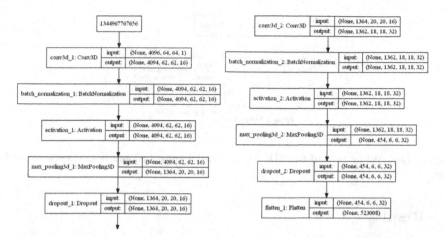

Fig. 4. Convolutional part of the 3D non-sequential convolutional neural network architecture

The second 3D CNN trained uses non-sequential architecture and is used to classify 4D fMRI images. Namely, after we convert the 4D image with resolution (H, W, L, T) into T 3D images with resolution (H, W, L), on each of the T 3D image we add a neuron which marks the position of the 3D image in the 4D image along the time axis. In this way we get two separate model parts: one convolutional part and one part that calculates the extra neuron vector representation. We concatenate the outputs of these two parts and pass them as input to two fully connected neural layers regularized by the dataset normalization technique. The final architecture can be seen in Fig. 3.

The convolutional part is shown on the left of Fig. 3. It consists of two convolutional layers with dimensions $(3, 3, 3, 1)$ filters, two MaxPooling layers and one output layer with resolution (H, W, C) into a vector with dimensions $H * W * C$. Convolutional layers are regularized by dataset normalization, and MaxPooling layers are dropout regularized. The architecture of the convolutional part of the non-sequential 3D CNN is shown in Fig. 4.

The part that calculates the vector representation of the extra neuron is shown in the right part of Fig. 3. It consists of a layer that embeds the neuron value in a 16-dimensional vector. Additionally, another layer is added which ensures that the output of the part that calculates the additional neuron vector representation is one-dimensional, so that it can be passed further through the fully connected part of the non-sequential 3D CNN. The part that calculates the additional neuron vector representation is not regularized. Its architecture can be seen in Fig. 5.

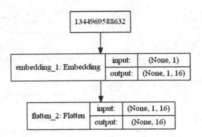

Fig. 5. Embedding part of the 3D non-sequential convolutional neural network architecture

4   Results

4.1   Experimental Setup

For standard machine learning models training, from the original data set we get two new data sets: a training set, which is composed of 80% of the data samples of the original data set, and a test set, which is composed of 20% of the data samples of the original data set, trying to maintain the class distribution in the entire data set. 10 fold-cross validation is used on the training set. After training, the set of hyperparameters that achieved the best average result in the 10 fold cross-validation process is selected. Finally, a final model is trained with the selected hyperparameters of the entire training data set and tested on the test data set.

For the CNN classification models, from the data set, by random selection without duplication, we select 20% of the data samples for the test set, again with retaining class distribution. From the remaining 80% data samples, by random selection without duplication, we select 20% of the data samples with class distribution maintenance, for the validation set. With this division, from the original data set we get three new data sets: training set, composed of 64% of the original data set, validation set, composed of 16% of the original data set, and a test set, of 20% of the original data set. When training neural networks, the results of the validation set are used to select the number of training epochs. In each epoch of neural network training, we calculate the value of the error function of the validation set. Using an early stopping technique, we stop training if currently the lowest value of the validation set error function has not been found in one of the last ten training epochs. The final model is the model that was trained in that epoch in which the value of the error function was the lowest.

To test the models we use five standard different metrics: accuracy, precision, sensitivity, F1 and AUC score.

4.2   T1 Images ASD Classification Results

Here we present the results obtained from the performed experiments. For each different type of images, the results achieved by each model are listed. T1 ASD

Table 1. T1 and fMRI images ASD classification results

Model	T1				fMRI			
	Accuracy	Precision	Sensitivity	F1	Accuracy	Precision	Sensitivity	F1
GNB	0.648	0.66	0.65	0.63	0.645	0.65	0.64	0.63
Ridge	0.591	0.59	0.59	0.59	0.659	0.66	0.66	0.66
DT	0.638	0.64	0.64	0.63	0.598	0.6	0.6	0.6
RF	0.679	0.68	0.68	0.67	0.681	0.67	0.67	0.67
ExT	**0.714**	0.7	0.7	0.7	**0.714**	0.7	0.7	0.7
GB	0.674	0.67	0.67	0.67	0.671	0.67	0.67	0.67
LR	0.597	0.59	0.6	0.59	0.671	0.67	0.67	0.67
SGD	0.592	0.63	0.59	0.58	0.593	0.59	0.59	0.59
MP	0.587	0.64	0.59	0.57	0.583	0.59	0.58	0.56
KNN	0.613	0.61	0.61	0.61	0.624	0.62	0.62	0.62
Lin SVM	0.582	0.65	0.58	0.49	0.655	0.69	0.65	0.63
2D CNN	0.679	0.68	0.68	0.68	0.644	0.64	0.64	0.64
3D CNN	0.648	0.65	0.65	0.64	0.671	0.68	0.67	0.66

classification is highest obtained using Extra Trees classifier. Random forests and CNNs also achieve good results, while the linear SVM, despite being a training dispute, achieves almost the worst results (see full results in Table 1).

4.3 fMRI Images ASD Classification Results

Experiments performed on fMRI achieve the highest average accuracy. Interesting to note is the consistency of this accuracy, which is simultaneously achieved with four different classifiers, two of which are parametric and two are non-parametric. Although the highest accuracy is not achieved with fMRI, such a consistency of experimental results and state-of-the-art average accuracy suggests that MRI ASD classification studies would probably be most successful if performed on functional magnetic resonance imaging. it is worth noting that even simple models, such as Gauss's naive Bayes classifier, achieve significantly higher results (see full results in Table 1).

Following are the ROC curves for the best models of the T1 and fMRI ASD classification. We note that the area under the ROC curve (AUC) is highest where the most successful experiments have been done with nonparametric models, classifiers using random forests and extra trees. Although not all nonparametric models, achieve the best results of accuracy, they still show that they are significantly better than the others when it comes to their ability to discriminate between the ASD patients and healthy control. The best ROC curves with AUC values can be seen in Fig. 6.

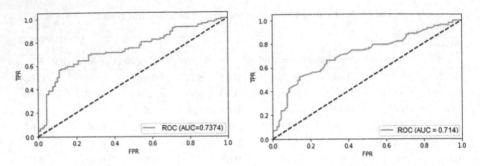

Fig. 6. Best T1 (left) and fMRI (right) ROC classification curve (with Extra trees classification model)

5   Conclusions

The objective in this research has been twofold: 1) to propose data transformation methods that deal with 3D structural T1 and 4D fMRI high dimensionality data; and 2) to explore ASD classification with standard machine learning algorithms and convolutional neural networks using the proposed data transformations. Our in-depth performance analysis have shown that even standard machine learning algorithms such as Random Forests, Extra Trees and Gradient Boosting with the proposed histograms calculation and data transformation method have achieved superior results to state-of-the-art classification on ABIDE II dataset. These classifiers give comparable results with the 2D and 3D CNNs proposed, although CNNs are expected to achieve significantly better results with MRI data [23]. This result is largely due to the low number of data samples available for this problem, which forces us to introduce a high degree of neural network regularization.

The result of 71.4% accuracy in classification, comparable with the state-of-the-art best literature result for ABIDE dataset [23], confirms there is a correlation between brain structure and ASD. Different types of MRI scans are likely to afford different opportunities for studying ASD as a neurodevelopmental disorder, although our air objective here was fMRI. We are currently focusing on novel data transformation and histograms calculation method with the use of the standard machine learning models using atlas-based regions of interest (ROI) histograms calculation; and to the CNNs models using different atlas-based ROI 3D-to-2D and 4D-to-3D transformation. We are expecting these models to be more precise, interpretable, furthermore, we plan to examine the relevance of each brain region to the classification itself. By doing this, the enhanced interpretable classification models can be used for other types of MRI imaging for classification of different neurodevelopmental diseases.

Acknowledgements. This work was partially financed by the Faculty of Computer Science and Engineering at the Ss. Cyril and Methodius University in Skopje.

References

1. Abadi, M., et al.: {TensorFlow}: a system for {Large-Scale} machine learning. In: 12th USENIX Symposium on Operating Systems Design and Implementation (OSDI 16), pp. 265–283 (2016)
2. Abraham, A., et al.: Deriving reproducible biomarkers from multi-site resting-state data: an autism-based example. Neuroimage **147**, 736–745 (2017)
3. Almuqhim, F., Saeed, F.: ASD-SAENet: a sparse autoencoder, and deep-neural network model for detecting autism spectrum disorder (ASD) using fMRI data. Front. Comput. Neurosci. **15**, 654315 (2021)
4. Artifact detection tools ART (2019). http://www.nitrc.org/projects/artifact_detect. Accessed 30 12 2019
5. Avants, B.B., Tustison, N., Song, G., et al.: Advanced normalization tools (ANTS). Insight J **2**(365), 1–35 (2009)
6. Batista, G., Silva, D.F., et al.: How k-nearest neighbor parameters affect its performance. In: Argentine Symposium on Artificial Intelligence, pp. 1–12. Citeseer (2009)
7. Bengs, M., Gessert, N., Schlaefer, A.: 4D Spatio-temporal deep learning with 4D fMRI data for autism spectrum disorder classification. arXiv preprint arXiv:2004.10165 (2020)
8. Bottou, L.: Large-scale machine learning with stochastic gradient descent. In: Lechevallier, Y., Saporta, G. (eds.) Proceedings of COMPSTAT 2010. Physica-Verlag HD, pp. 177–186. Springer (2010). https://doi.org/10.1007/978-3-7908-2604-3_16
9. Chollet, F., et al.: Keras: The python deep learning library. Astrophy. Source Code Libr., pp. ascl-1806 (2018)
10. Di Martino, A., et al.: Enhancing studies of the connectome in autism using the autism brain imaging data exchange II. Sci. data **4**(1), 1–15 (2017)
11. DI Martino, A., et al.: The autism brain imaging data exchange: towards a large-scale evaluation of the intrinsic brain architecture in autism. Mol. Psychiatry **19**(6), 659–667 (2014)
12. Dougherty, D.D., Rauch, S.L., Rosenbaum, J.F.: Essentials of Neuroimaging for Clinical Practice. American Psychiatric Pub, Washington (2008)
13. El Gazzar, A., Cerliani, L., van Wingen, G., Thomas, R.M.: Simple 1-D convolutional networks for resting-state fMRI based classification in autism. In: 2019 International Joint Conference on Neural Networks (IJCNN), pp. 1–6. IEEE (2019)
14. El-Gazzar, A., Quaak, M., Cerliani, L., Bloem, P., van Wingen, G., Mani Thomas, R.: A hybrid 3DCNN and 3DC-LSTM based model for 4D spatio-temporal fMRI data: an ABIDE autism classification study. In: Zhou, L., et al. (eds.) OR 2.0/MLCN -2019. LNCS, vol. 11796, pp. 95–102. Springer, Cham (2019). https://doi.org/10.1007/978-3-030-32695-1_11
15. Fan, R.E., Chang, K.W., Hsieh, C.J., Wang, X.R., Lin, C.J.: LIBLINEAR: a library for large linear classification. J. Mach. Learn. Res. **9**, 1871–1874 (2008)
16. Friedman, J.H.: Stochastic gradient boosting. Comput. Stat. Data Anal. **38**(4), 367–378 (2002)
17. Geurts, P., Ernst, D., Wehenkel, L.: Extremely randomized trees. Mach. Learn. **63**(1), 3–42 (2006). https://doi.org/10.1007/s10994-006-6226-1
18. Hoerl, A.E., Kannard, R.W., Baldwin, K.F.: Ridge regression: some simulations. Commun. Stat. Theory Methods 4(2), 105–123 (1975)

19. Hosmer, D.W., Jr., Lemeshow, S., Sturdivant, R.X.: Applied Logistic Regression, vol. 398. Wiley, Hoboken (2013)
20. Ioffe, S., Szegedy, C.: Batch normalization: accelerating deep network training by reducing internal covariate shift. In: International Conference on Machine Learning, pp. 448–456. PMLR (2015)
21. Jenkinson, M., Beckmann, C.F., Behrens, T.E., Woolrich, M.W., Smith, S.M.: FSL. Neuroimage **62**(2), 782–790 (2012)
22. Jiang, H., Cao, P., Xu, M., Yang, J., Zaiane, O.: Hi-GCN: a hierarchical graph convolution network for graph embedding learning of brain network and brain disorders prediction. Comput. Biol. Med. **127**, 104096 (2020)
23. Khodatars, M., et al.: Deep learning for neuroimaging-based diagnosis and rehabilitation of autism spectrum disorder: a review. Comput. Biol. Med. **139**, 104949 (2021)
24. Khosla, M., Jamison, K., Kuceyeski, A., Sabuncu, M.R.: 3D convolutional neural networks for classification of functional connectomes. In: Stoyanov, D., et al. (eds.) DLMIA/ML-CDS -2018. LNCS, vol. 11045, pp. 137–145. Springer, Cham (2018). https://doi.org/10.1007/978-3-030-00889-5_16
25. Kim, B.H., Ye, J.C.: Understanding graph isomorphism network for rs-fMRI functional connectivity analysis. Front. Neurosci., 630 (2020)
26. Kingma, D.P., Ba, J.: ADAM: a method for stochastic optimization. arXiv preprint arXiv:1412.6980 (2014)
27. Li, X., Morgan, P.S., Ashburner, J., Smith, J., Rorden, C.: The first step for neuroimaging data analysis: DICOM to NIfTI conversion. J. Neurosci. Methods **264**, 47–56 (2016)
28. Li, X., et al.: BrainGNN: interpretable brain graph neural network for fMRI analysis. Med. Image Anal. **74**, 102233 (2021)
29. Liaw, A., Wiener, M., et al.: Classification and regression by randomForest. R News **2**(3), 18–22 (2002)
30. Pedregosa, F., et al.: Scikit-learn: machine learning in python. J. Mach. Learn. Res. **12**, 2825–2830 (2011)
31. Qayyum, A., et al.: An efficient 1DCNN-LSTM deep learning model for assessment and classification of fMRI-based autism spectrum disorder. In: Raj, J.S., Kamel, K., Lafata, P. (eds.) Innovative Data Communication Technologies and Application, vol. 96, pp. 1039–1048. Springer, Singapore (2022). https://doi.org/10.1007/978-981-16-7167-8_77
32. Riaz, A., Asad, M., Alonso, E., Slabaugh, G.: DeepFMRI: End-to-end deep learning for functional connectivity and classification of ADHD using fMRI. J. Neurosci. Methods **335**, 108506 (2020)
33. Riedmiller, M.: Advanced supervised learning in multi-layer perceptrons-from backpropagation to adaptive learning algorithms. Comput. Stan. Interfaces **16**(3), 265–278 (1994)
34. Rish, I., et al.: An empirical study of the naive Bayes classifier. In: IJCAI 2001 Workshop on Empirical Methods in Artificial Intelligence, vol. 3, pp. 41–46 (2001)
35. Safavian, S.R., Landgrebe, D.: A survey of decision tree classifier methodology. IEEE Trans. Syst. Man Cybern. **21**(3), 660–674 (1991)
36. Sarraf, S., Desouza, D.D., Anderson, J.A., Saverino, C.: MCADNNeT: recognizing stages of cognitive impairment through efficient convolutional fMRI and MRI neural network topology models. IEEE Access **7**, 155584–155600 (2019)
37. Serai, S.D.: Basics of magnetic resonance imaging and quantitative parameters T1, T2, T2*, T1rho and diffusion-weighted imaging. Pediatr. Radiol. **52**(2), 217–227 (2021). https://doi.org/10.1007/s00247-021-05042-7

38. Smith, S.M.: Bet: brain extraction tool. FMRIB TR00SMS2b, Oxford Centre for Functional Magnetic Resonance Imaging of the Brain), Department of Clinical Neurology, Oxford University, John Radcliffe Hospital, Headington, UK (2000)
39. Statistical parametric mapping SPM12 (2018). https://www.fil.ion.ucl.ac.uk/spm/software/spm12/. Accessed 30 12 2019
40. Sporns, O.: Structure and function of complex brain networks. Dialogues Clin. Neurosci. **15**, 247–262 (2022)
41. Srivastava, N., Hinton, G., Krizhevsky, A., Sutskever, I., Salakhutdinov, R.: Dropout: a simple way to prevent neural networks from overfitting. J. Mach. Learn. Res. **15**(1), 1929–1958 (2014)
42. Tahmassebi, A., Gandomi, A.H., McCann, I., Schulte, M.H., Goudriaan, A.E., Meyer-Baese, A.: Deep learning in medical imaging: fMRI big data analysis via convolutional neural networks. In: Proceedings of the Practice and Experience on Advanced Research Computing, pp. 1–4 (2018)
43. Thomas, R.M., Gallo, S., Cerliani, L., Zhutovsky, P., El-Gazzar, A., Van Wingen, G.: Classifying autism spectrum disorder using the temporal statistics of resting-state functional MRI data with 3D convolutional neural networks. Front. Psych. **11**, 440 (2020)
44. Tournier, J.D., et al.: Mrtrix3: a fast, flexible and open software framework for medical image processing and visualisation. Neuroimage **202**, 116137 (2019)
45. Vieira, S., Pinaya, W.H., Mechelli, A.: Using deep learning to investigate the neuroimaging correlates of psychiatric and neurological disorders: methods and applications. Neurosci. Biobehav. Rev. **74**, 58–75 (2017)
46. Wang, L.: Support Vector Machines: Theory and Applications, vol. 177. Springer Science & Business Media, Berlin (2005)
47. Wen, D., Wei, Z., Zhou, Y., Li, G., Zhang, X., Han, W.: Deep learning methods to process fMRI data and their application in the diagnosis of cognitive impairment: a brief overview and our opinion. Front. Neuroinform. **12**, 23 (2018)
48. Wen, J., et al.: Convolutional neural networks for classification of Alzheimer's disease: Overview and reproducible evaluation. Med. Image Anal. **63**, 101694 (2020)
49. Whitfield-Gabrieli, S., Nieto-Castanon, A.: CONN: a functional connectivity toolbox for correlated and anticorrelated brain networks. Brain Connect. **2**, 125–41 (2012). https://doi.org/10.1089/brain.2012.0073
50. Woolrich, M.W., et al.: Bayesian analysis of neuroimaging data in FSL. Neuroimage **45**(1), S173–S186 (2009)
51. Yang, X., Zhang, N., Schrader, P.: A study of brain networks for autism spectrum disorder classification using resting-state functional connectivity. Mach. Learn. Appl. **8**, 100290 (2022)
52. Yin, W., Li, L., Wu, F.X.: Deep learning for brain disorder diagnosis based on fMRI images. Neurocomputing **469**, 332–345 (2022)

Detection of High Noise Levels in Electrocardiograms

Danche Papuchieva$^{(\boxtimes)}$ and Marjan Gusev

Ss Cyril and Methodius University in Skopje, Skopje, North Macedonia
dpapucieva@gmail.com, marjan.gushev@finki.ukim.mk
https://www.ukim.edu.mk/

Abstract. Wearable electrocardiogram sensor technology trends show increased use by patients in their homes and offices. However, numerous issues arise with physically active patients since physical movements and muscle noise generate artifacts and corrupt the signal to high levels resulting in false detection of heartbeats and arrhythmia. Therefore, this introduces the need for an algorithm that detects if the collected data represent a regular electrocardiogram or if it is a noisy signal. The detection of regions with high noise levels, which make the signal uninterpretable, is realized by signal processing methods and applying different digital filters. In this paper, we aim to prove a research hypothesis if the variance is not (less) sensitive to heart rate versus energy for noise detection in the electrocardiogram. In addition, the research questions analyzed in this paper address finding optimal values for threshold, window length, and offset, and determining the dependence on heart rate, beat type, and peak height. We evaluate the enhanced MIT-BIH arrhythmia benchmark database by adding noise to selected time intervals. The results show that the variance-based method outperforms the energy-based method, mainly due to the independence of window length and heart rate. However, both methods are affected by the QRS peak amplitude.

Keywords: ECG · QRS · Noise detection · Signal processing

1 Introduction

An electrocardiogram (ECG) is an electrical representation of the heart's activity. Physical and muscle movements generate artifacts and corrupt the ECG signal to high levels making a wrong detection of heartbeats and associated arrhythmia. Algorithm improvement for high noise presence detection leads to a better performance algorithm for heartbeats and arrhythmia detection.

The level of noise present in the ECG can be measured by the Signal-To-Noise ratio (SNR), usually expressed in decibels (dB). We have tested our algorithm [3] and found that levels up to $\text{SNR} \leq -12$ dB can be detected with high performance [1], and when the noise level increases over $\text{SNR} \geq -6$ dB, performance decreases a lot.

© The Author(s), under exclusive license to Springer Nature Switzerland AG 2022
K. Zdravkova and L. Basnarkov (Eds.): ICT Innovations 2022, CCIS 1740, pp. 190–204, 2022.
https://doi.org/10.1007/978-3-031-22792-9_15

In this research, we analyze two signal processing methods to detect high noise levels in the ECG, the first one based on the energy, as reported in our earlier paper [8], and the second one on the variance. We set a research hypothesis that the variance-based method performs better than the energy-based method. These methods calculate a corresponding value for a sliding window with predefined width, and the decision is made by comparing this value versus a threshold. For these research questions, we aim to find the optimal window length, sliding window offset, and the corresponding threshold to determine if the analyzed ECG consists of a clean ECG signal or is corrupted by noise. In addition, we aim to find if the variance is not sensitive to the heart rate versus energy, and if the energy calculated on a window with a given length is more prone to changes than the variance. The number of heartbeats in one window varies along with the heart rate, and thus the energy changes, making the energy-based algorithm sensitive to the heart rate.

The width of the sliding window also plays a crucial role in the sensitivity of noise detection when the variance approach is used. For example, a shorter window introduces higher variance in case of abrupt changes, which leads to false positive detections (noise detected where noise does not exist). The origin of these abrupt changes may be cardiac activity, such as ventricular beats.

We conducted several experiments to prove the research hypothesis and investigate the research questions. The conducted experiments aim to find the optimal window length and determine the overall variance threshold for the variance-based method since the results of the energy-based method are already reported in [8]. The used ECG benchmark databases are the MIT-BIH Noise Stress Test database and the MIT-BIH Arrhythmia database, which were used to generate several datasets adding noise regions with different noise levels to approach the problem with a higher number of ECG datasets. The evaluation is realized according to the duration-based methods, where positives address the regions with noise-corrupted ECG, and negatives are the regions with regular ECG signals.

The rest of the paper is organized according to the following structure. Section 2 presents the related works. The experimental research methods are explained in Sect. 3 and results presented in Sect. 4. Section 5 is devoted to evaluation and discussion of obtained results, and Sect. 6 to conclusions and future work directions.

2 Related Works

Motion artifacts inspired a lot of researchers to find a way to eliminate them, as elaborated in a comparative review [5]. An approach is introduced in [7] using the electrode-skin impedance as a second channel to reduce motion artifacts. Discrete wavelet transform is applied to delete motion artifacts that are time and frequency selective. The problem of removing noise and artifacts from ECGs is tested on 3-channel ECG recordings taken from human subjects in [2] revealing results that the applied Independent Component Analysis method can detect and remove a variety of noise and artifact sources in these ECGs.

Better-quality ECGs could be reconstructed by applying a novel ECG denoising technique based on the variable frequency complex de-modulation algorithm [6] to address noises from a variety of sources with sub-band decomposition.

Related work is also elaborated in our earlier paper [8], analyzing the agglomerative clustering approach [13] to the problem of artifact detection in ECG signals, a two-step detection method [9] with first-order intrinsic mode function and use of three predefined thresholds: Shannon entropy, mean and variance. A robust solution for detection of ECG artifacts [15] using Fourier transform, a solution based on decision trees [10], and a model with autocorrelation [15] are examples of other related approaches, but without practical significance in a real-time environment.

Note that there is much research on lower noise level elimination, especially those imposed by the baseline wander or high-level noise, which can be easily eliminated correspondingly by high-pass and low-pass digital filters, and this is not the intention of this research. We aim to detect uninterpretable noisy ECG signals, and the noise is detected to exclude these corrupted segments from further ECG analysis.

3   Methods

This section introduces the dataset, variance and energy-based algorithm approaches, experimental setup, and evaluation methodology.

3.1   Dataset

We started analyzing the MIT-BIH Noise Stress Test Database (NST) [12], which contains two records, 118 and 119, in six versions, each with different noise levels (SNR = 24 dB, 18 dB, 12 dB, 6 dB, 0 dB, and −6 dB).

Note that SNR =0 dB means the signal and noise levels are equal, and negative values indicate a noise level higher than the signal level.

For deeper analysis, we created an extension of the entire reference MIT-BIH Arrhythmia ECG database (MITDB) [11]. This database contains 48 half-hour ECG recordings, of which only 44 records are used since the others are generated by patients with pacemakers. Our extensions to the MITDB are created by adding calibrated amounts of noise to the ECG recordings with different SNR levels. The noise recordings are made with the help of physically active volunteers and standard ECG recorders, leads, and electrodes. The electrodes were placed on the limbs in positions in which the subjects' ECGs were not visible [4, 12].

The three noise records are assembled from the recordings by selecting intervals that predominantly contain baseline wander, muscle (EMG) artifact, and electrode motion artifact. Electrode motion artifact is generally considered the most troublesome since it can mimic the appearance of ectopic beats and cannot be easily removed with simple filters like other types of noise.

Since the original ECG recordings are clean (free of noise) and contain regular ECG signals, the correct beat annotations are known even when the noise makes

Table 1. Datasets with ECG signals corrupted with noise

Dataset	SNR (dB)
MITDBe_6	-6
MITDBe_4	-4
MITDBe_2	-2
MITDBe00	0
MITDBe02	2
MITDBe04	4
MITDBe06	6
MITDBe12	12
MITDBe18	18
MITDBe24	24

the recordings visually unreadable. The reference annotations for these records are simply copies of those for the original clean ECGs.

Noise is added to the original ECG samples after the first 5 min of each record, during two-minute segments alternating with two-minute clean intervals. Our experimental datasets with additional noise added to MITDB records with corresponding signal-to-noise ratios (SNRs) during the noisy segments are presented in Table 1.

The original MITDB database is sampled 360 Hz with a 11-bit resolution. For this research, the recordings are resampled 125 Hz, with a 12-bit resolution, to cope with larger amplitudes for low SNR levels.

3.2   Algorithm

In this research, the variance characteristic is observed in clean and noisy segments of ECG recordings and compared with the noise detection algorithm based on energy calculation from previous related work on this topic [1].

The variance and energy are calculated on the ECG data without prior data preprocessing. The ECG signal is only normalized before the analysis by (1), where $nbit$ is the bit resolution.

$$ecg_{normalised}[n] = \frac{ecg[n] - 2^{(nbit-1)}}{2^{(nbit-1)}} \tag{1}$$

Calculation of variance is according to the Eq. (2), where x_i is each sample value, μ is the mean value of window length signal fragment, and n is the number of samples per window.

$$variance = \frac{\sum_{i=1}^{n}(x_i - \mu)^2}{n - 1} \tag{2}$$

The energy is calculated with the differential filter from previous related work. The differential filter is defined by the Eq. (3), where $(x[i] - x[i-1])$ is the difference between two consecutive samples.

$$\sum_{i=1}^{n}(x[i] - x[i-1])$$ (3)

3.3   Experimental Setup

The experiment introduces a sliding window method with a fixed sliding window length. To determine the optimal window length, the experiment is defined by test cases with window lengths from 2.5 to 15 s with a step of 0.5 s.

To achieve more accurate detection of the segment where the noise starts, we analyze different window displacements (shifts). The window shifting should be shorter when higher precision is required to estimate the beginning and end of the segment. For this purpose, the variance and energy are observed with 1, 10, 20, 30, 50, and 100 samples window shift.

To determine the optimal threshold for the noise level, evaluation of the noise detection algorithm is performed with several variance thresholds: 100e−6, 200e−6, 300e−6, 400e−6, and 500e−6. These thresholds were selected by prior visual observation (Fig. 1).

Analyzing the domain space of the test cases with 25 different values of window length, 6 values of window offset, 5 threshold values, 10 different datasets, and 44 records per dataset, we conducted a total of 330.000 tests.

3.4   Evaluation Methods

True Positive (TP) is an outcome where the model correctly predicts actual noise presence, and False Positive (FP), an outcome where the model incorrectly predicts noise in intervals without noise in the signal. The opposite applies to true negative (TN) and false negative (FN). If the outcome is negative and noise does not exist - TN, and if the outcome is negative and the signal is noise-free - FN.

A confusion matrix is formed from the four outcomes produced as a result of binary classification. Various measures can be derived from a confusion matrix. In this paper, the most relevant metric is the F1 score, which is a harmonic mean of *Precision*, also known as positive predictive value $(PPV = TP/(TP + FP))$ and *Recall*, also known as sensitivity $(SEN = TP/(TP + FN))$ [14].

The evaluation is performed by calculating the best F1 score for each signal using the duration method. Each sample in clean and noisy segments is evaluated, and the F1 score is calculated by (4) based on the prediction of whether the analyzed segment is TP, TN, FP, or FN.

$$F1 = \frac{2 * PPV * SEN}{PPV + SEN}$$ (4)

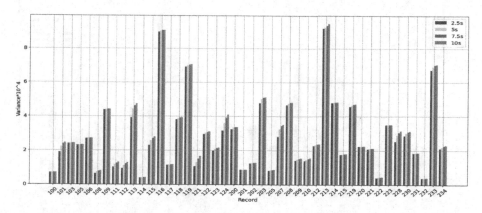

Fig. 1. Average **variance** calculated for MITDBe24 with different window lengths

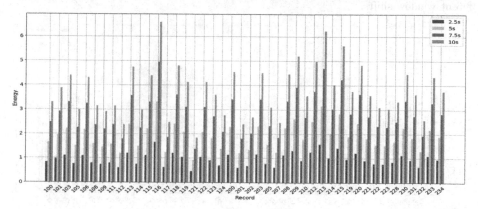

Fig. 2. Average **energy** calculated for MITDBe24 with different window lengths

4   Results

Next, we present achieved results organized to address their dependence on window length and shift, heart rate, and threshold.

4.1   Dependence on Window Length and Heart Rate

Figures 1 and 2 present the average variance and energy level for signal noise levels where the ECG signal is dominant (SNR = 24 dB). We present only four values of window length, although all the experiments were conducted for all test cases with window lengths from 2.5 to 15 s for all datasets. The results show that the variance is not influenced by the window length, while the energy is linearly proportional to the window length.

By visual observation, it can be concluded that a predefined fixed threshold cannot achieve high accuracy for noise detection. The noise level varies in each recording and is influenced by the ECG.

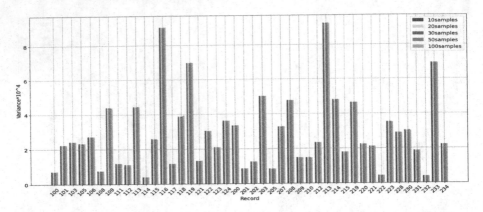

Fig. 3. Average **variance** calculated for MITDBe24 with 5 s window length and different window shifts.

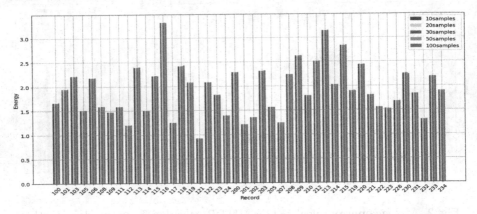

Fig. 4. Average **energy** calculated for MITDBe24 with 5 s window length and different window shifts.

The variance value changes only in small increments, opposite to the energy, which is proportional to the window length and thus to heart rate.

The conclusion related to the window length is specified by the algorithm accuracy after the evaluation.

4.2   Influence of the Window Shift

We observed that the window shift (offset) has no or small negligible impact on the variance and energy distribution. It only affects the resolution. This is shown in Figs. 3 and 4. The presented variance and energy are calculated on a 5-s sliding window with 10, 20, 30, 50, and 100 samples shift.

Table 2. Average F1 score for noise detection with variance on different datasets for predefined threshold with window length 5 s, and window shift 10 samples

Dataset	F1 (%)
MITDBe_6	92.66
MITDBe_4	86.54
MITDBe_2	87.99
MITDBe00	91.29
MITDBe02	88.62
MITDBe04	86.16
MITDBe06	80.47
MITDBe12	52.97
MITDBe18	27.15
MITDBe24	16.37

4.3 Threshold

The impact of the window lengths and the window shifts are evaluated for different threshold levels. Based on the F1 score, the best-predefined threshold is determined for each subject and each dataset, determined by the corresponding SNR level.

Table 2 presents the overall F1 score calculated for all datasets applying the variance approach with the predefined threshold for a window length of 5 s and offset of 10.

Figure 5 shows the dependence of the overall F1 score on a different dataset with a corresponding SNR level. Each line represents the F1 score for several tested thresholds. The x-axis presents the SNR level, and we can observe how the F1 score decreases when the noise level increase. Observing Fig. 5, it can be concluded that the length of window does not affect the variance as much, and the best results are obtained for window lengths of 5 and 7.5 s.

Even though high accuracy can be achieved independently for each subject, it is difficult to obtain an overall predefined threshold that will perform with high accuracy for different subjects and SNR levels. If the amplitude of the QRS complex is higher, which is typical for persons with a smaller amount of fat below the skin and better contact with the ECG electrode, the F1 score on the same threshold decreases rapidly. This is observed in Fig. 6, where the same evaluation is done on the signals, with scaled amplitude twice, and keeping the same thresholds for the variance.

The results of the hard-coded threshold, lead to the conclusion that adaptive threshold method is most suitable for this type of signals.

A lower F1 score than expected is achieved for $SNR = -4$ dB and $SNR = -2$ dB because the data are damaged in several segments (due to the conversion of bit resolution, cut-off values of the composed signal, and noise over the bit-resolution maximal value).

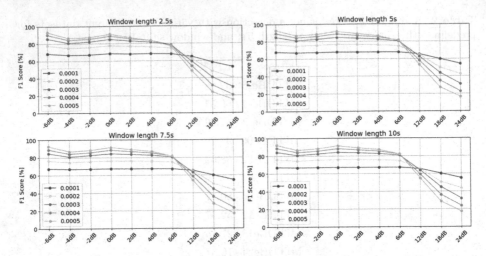

Fig. 5. Overall F1 score for different thresholds and window lengths

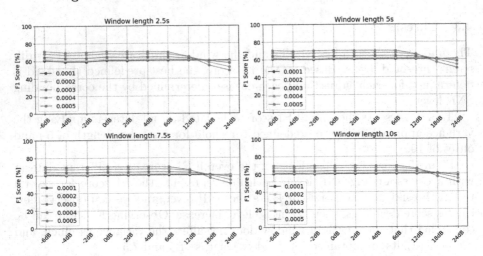

Fig. 6. Overall F1 score for different thresholds and window lengths for higher signal amplitude

5   Discussion

This section is a broad discussion regarding the conclusions gathered during the analysis. Results are discussed to express the influence of the heart rate, QRS peak height, optimal threshold selection, window size, and shift.

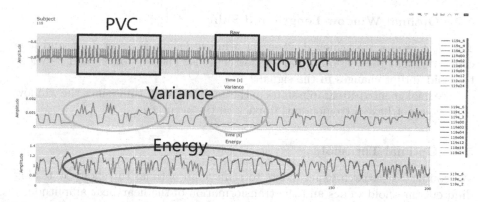

Fig. 7. Variance and energy distribution for MITDB record 119 calculated with window length 2.5 s, window shift 100 samples.

5.1 Dependence on Heart Rate

We observed that the variance is not affected by the window length. Therefore, the variance is also not affected if there are more heartbeats in the analyzed window, assuming there is at least one heartbeat in the analyzed window. This leads to the conclusion that the variance is not affected by the heart rate.

Otherwise, the energy is proportionally dependent on the window length. The higher the heart rate, the higher number of heartbeats found in the analyzed window. Since the energy is directly dependent on the number of heartbeats in the analyzed window, therefore, the energy is influenced by the heart rate, as opposed to the variance.

5.2 Dependence on Heartbeat Type and Peak Height

Heartbeat type influences both the level of energy and variance. For example, ventricular beats result with higher energy and much higher variance than normal heartbeats.

Figure 7 illustrates a part of the MITDB record 119, which contains a lot of ventricular heartbeats. The first row represents the ECG, the second associated variance, and the third is the energy. Note that the variance and energy are calculated by the sliding window approach using a window length of 2.5 s and a window shift of 100 samples. We observe that the variance in the regions without ventricular beats is much lower than in regions containing ventricular beats. Although the presence of ventricular beats affects the energy, the impact is not overexposed.

In addition, we observe that the heartbeat peak height influences the energy level and variance. The higher the heartbeat peak is, the higher the variance and energy are. This conclusion affects the determination of the threshold for noise detection.

5.3 Optimal Window Length and Shift

The best results are achieved when the window length is 5 s or 7.5 s, and it is concluded that these are optimal window lengths. The window should be wider to avoid abrupt changes in the signal.

We implemented 10 sample shifts in the analyzed algorithms since the window shift does not have a tremendous impact on the accuracy.

5.4 Optimal Threshold

We conducted research experiments to achieve an optimal threshold by testing different threshold values and the transformation of the heartbeat amplitude.

Predefined Threshold. Evaluating the achieved performance for the detection of high-level noise, we observe various behavior of different records from the given dataset, which leads to the conclusion that there is no unique threshold value to achieve the best performance, which is influenced by the beat type and the number of heartbeats present in the analyzed window (indirectly on the heart rate).

It is extremely difficult to create a general robust noise detection model with a predefined threshold, especially in the cases of different peak-to-peak heartbeat amplitudes, which depend on the fat level under the skin, skin conductivity of electrical signal, and ECG electrode contact quality.

ECG measured on different individuals shows that the heartbeat amplitude is usually around 1 mV, but it may often be higher up to 2 mV in younger and lower up to 0.6 mV in obese individuals.

Adaptive Threshold. The variance and the energy will vary proportionally to the heartbeat peak amplitude.

A good approach to cope with the problem of the different amplitudes of the heartbeat peak is to use an adaptive threshold.

One idea for adaptive threshold is to specify the threshold according to heartbeat peak amplitude. We propose two methods for the calculation of the adaptive threshold:

- **Min/Max** method that determines the threshold based on min and max values in the analyzed window with a clean ECG signal,
- **Energy**-based method that determines the threshold based on energy calculated in the analyzed window with a clean ECG signal that contains two heartbeats,

The Min/Max method calculates the minimum and maximum of an analyzed window with a length of 2.5 s. The threshold is calculated at the beginning of the recording when the signal is clean of noise. In a real environment, it will be a problem to conclude whether the signal is clean ECG or corrupted by noise. This prior knowledge makes the calculation of an adaptive threshold problematic.

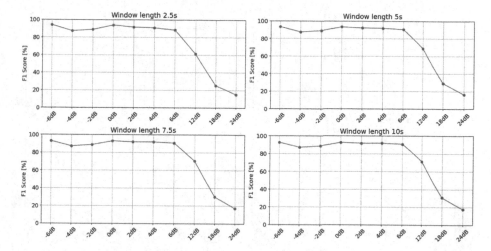

Fig. 8. Achieved overall F1 score on MITDB using an adaptive threshold calculated with the min/max approach

This approach achieves good overall results for SNR 6 dB, as presented in Fig. 8 and Table 3.

The energy-based approach is based on calculating the energy in a window segment that consists of two heartbeats. This knowledge of the existence of two heartbeats in the first analyzed window to determine the adaptive threshold is once again problematic.

The algorithm performance with this approach is worse than the previous min/max approach to determine the threshold. This approach depends on the heart rate, and in this research, the average heart rate is calculated during the entire recording.

Higher accuracy for noise detection is expected in a real-time scenario when the heart rate is estimated by other methods.

Figure 9 and Table 3 present the achieved results for the energy-based method. Figure 10 presents the results for the adaptive threshold when QRS amplitude is higher. The adaptive threshold can be further optimized to achieve better results for variable QRS amplitude.

5.5   Comparison to Other Results

Our research on related work has not found any paper on detecting whether the signal can be treated as noise or regular ECG, besides our earlier work [8]. All other researchers have focused on detecting lower noise levels and removing them from the ECG, but not detecting intervals that cannot be interpreted.

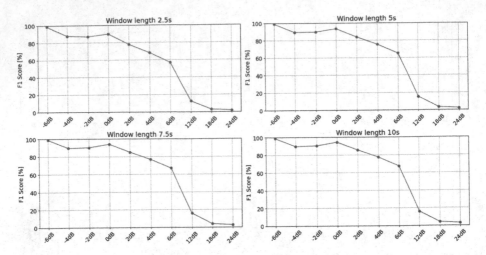

Fig. 9. Achieved overall F1 score on MITDB using an adaptive threshold calculated with the energy-based method

Table 3. Average F1 score on MITDB based on determination of an adaptive threshold with two different methods

Dataset	Min/Max	Energy
MITDBe-6	93.29	98.73
MITDBe-4	87.29	89.24
MITDBe-2	88.87	89.55
MITDBe00	93.20	93.26
MITDBe02	92.00	83.58
MITDBe04	91.83	75.16
MITDBe06	90.38	64.94
MITDBe12	68.73	15.22
MITDBe18	29.10	03.40
MITDBe24	16.33	02.25

6   Conclusion

In this research, we have introduced a variance-based method and compared it to the energy-based method to detect high noise levels in the ECG where the signal becomes uninterpretable. Our target was to find where each method is applicable and which does not depend on heart rate or signal amplitude. Finally, we conducted numerous experiments to find the optimal window length, shift, and threshold level and decide whether the analyzed window is an interpretable ECG signal.

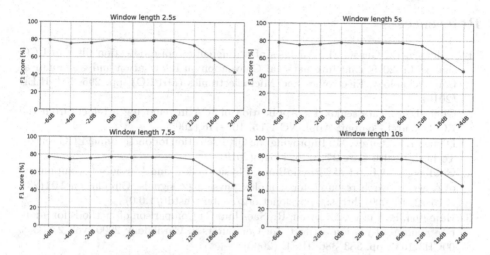

Fig. 10. F1 score - Adaptive threshold with higher QRS amplitude evaluated with Min/Max method

We have proved the hypothesis that variance is not (less) sensitive to heart rate versus energy by observing the performance of noise detection methods at various window lengths. Both the variance and energy-based methods are dependent on the peak amplitude. Prior knowledge of the amplitude is the essence of both algorithms and prevents the usage of a predefined threshold value.

According to the results, the best results are achieved when the window length is 5 s or 7.5 s, and due to the lower complexity of calculations, we recommend a window with a 5 s length. Since the window shift does not have a tremendous impact on the accuracy, 10 sample shifting is acceptable as the most convenient.

Achieving high accuracy with 500e−6 predefined thresholds for noise levels where SNR ≤ 6 dB, and even better results with adaptive threshold, promises an optimized model with an adaptive threshold for variable QRS peak-to-peak amplitude, which will be implemented in an automated pipeline for ECG analysis.

Regarding the adaptive threshold, our future work will address the distribution of clean and noisy signals in more detail. Finding the dependence of the threshold on the QRS peak amplitude is the essence of any further analysis. A big challenge is to detect high noise levels without prior knowledge of the peak amplitude. The use of any additional parameter will increase the algorithm complexity, and it is a challenge to adjust the appropriate combination of thresholds for each parameter. Therefore, a simple ML model can be developed to determine the optimal adaptive threshold using previously investigated features (variance and energy) and additional features such as skewness, kurtosis, and mean crossing. Finally, this research leads toward finding an algorithm for analyzing time series data and concluding whether they represent an ECG signal.

References

1. Ajdaraga, E., Gusev, M.: Analysis of a differential noise detection filter in ECG signals. In: 2019 42nd International Convention on Information and Communication Technology, Electronics and Microelectronics (MIPRO), pp. 295–300. IEEE (2019)
2. Clifford, G., Tarassenko, L.: Application of ICA in removing artefacts from the ECG. Neural Comput. Appl. NCA **15**, 105–116 (2005)
3. Domazet, E., Gusev, M.: Improving the QRS detection for one-channel ECG sensor. Technol. Health Care **27**(6), 623–642 (2019)
4. Goldberger, A.L., et al.: PhysioBank, PhysioToolkit, and PhysioNet: components of a new research resource for complex physiologic signals. Circulation **101**(23), e215–e220 (2000). https://physionet.org/content/nstdb/1.0.0/
5. Hamilton, P., Curley, M., Aimi, R., Sae-Hau, C.: Comparison of methods for adaptive removal of motion artifact. In: Computers in Cardiology 2000, vol. 27 (Cat. 00CH37163), pp. 383–386. IEEE (2000)
6. Hossain, M.B., Bashar, S.K., Lazaro, J., Reljin, N., Noh, Y.S., Chon, K.: A robust ECG denoising technique using variable frequency complex demodulation. Comput. methods programs biomed. **200**, 105856 (2020). https://doi.org/10.1016/j.cmpb.2020.105856
7. Kirst, M., Glauner, B., Ottenbacher, J.: Using DWT for ECG motion artifact reduction with noise-correlating signals. In: 2011 Annual International Conference of the IEEE Engineering in Medicine and Biology Society, pp. 4804–4807 (2011). https://doi.org/10.1109/IEMBS.2011.6091190
8. Krluku, E.A., Gusev, M.: Detection of uninterpretable ECG signal segments. In: 2020 43rd Information Convention on Information, Communication and Electronic Technology (MIPRO), pp. 337–342. IEEE (2020)
9. Lee, J., McManus, D.D., Merchant, S., Chon, K.H.: Automatic motion and noise artifact detection in Holter ECG data using empirical mode decomposition and statistical approaches. IEEE Trans. Biomed. Eng. **59**(6), 1499–1506 (2011)
10. Moeyersons, J., Varon, C., Testelmans, D., Buyse, B., Van Huffel, S.: ECG artefact detection using ensemble decision trees. In: 2017 Computing in Cardiology (CinC), pp. 1–4. IEEE (2017)
11. Moody, G.B., Mark, R.G.: The impact of the MIT-BIH arrhythmia database. IEEE Eng. Med. Biol. Mag. **20**(3), 45–50 (2001)
12. Moody, G.B., Muldrow, W., Mark, R.G.: A noise stress test for arrhythmia detectors. Comput. Cardiol. **11**(3), 381–384 (1984)
13. Rodrigues, J., Belo, D., Gamboa, H.: Noise detection on ECG based on agglomerative clustering of morphological features. Comput. Biol. Med. **87**, 322–334 (2017)
14. Saito, T., Rehmsmeier, M.: Basic evaluation measures from the confusion matrix (2016). https://classeval.wordpress.com/introduction/basic-evaluation-measures/
15. Varon, C., Testelmans, D., Buyse, B., Suykens, J.A., Van Huffel, S.: Robust artefact detection in long-term ECG recordings based on autocorrelation function similarity and percentile analysis. In: 2012 Annual International Conference of the IEEE Engineering in Medicine and Biology Society, pp. 3151–3154. IEEE (2012)

Author Index

Alcaraz, Salvador 38
Andonov, Stefan 13
Andova, Ivona 121
Angelovski, Gorast 76

Bajrami, Merxhan 121
Bernad, Cristina 38
Bogdanova, Ana Madevska 161

Chitkushev, Ljubomir 76

Damaševičius, Robertas 63
Demukaj, Venera 147
Dimitrova, Vesna 13
Dimitrovski, Ivica 107
Dineva, Katarina Trojachanec 107, 121
Dobreva, Jovana 13, 51

Ferati, Mexhid 147
Filiposka, Sonja 38

Gievska, Sonja 175
Gilly, Katja 38
Gusev, Marjan 190

Ilievska, Natasha 26
Ilijoski, Bojan 121
Ivanoska, Ilinka 175

Jankova, Dona 121

Kalajdziski, Slobodan 175
Kapočiūtė-Dzikienė, Jurgita 63
Kitanovski, Ivan 107
Kjorveziroski, Vojdan 38
Kocarev, Ljupcho 51

Krajevski, Jovan 175
Kurti, Arianit 147

Lameski, Petre 121
Loshkovska, Suzana 107

Marojevikj, Jovana 76
Mihajloska, Hristina 13
Mishev, Kostadin 51
Mörtberg, Christina 147

Papuchieva, Danche 190
Pavlov, Tashko 51
Pejov, Ljupcho 161
Peshov, Hristijan 76
Popovska-Mitrovikj, Aleksandra 13

Roig, Pedro Juan 38
Rusevski, Ivan 76

Sandjakoska, Ljubinka 161
Simjanoska, Monika 51
Spirovska, Eva 76
Stefanovska, Emilija 93

Tasevski, Ivo 13
Tesfagergish, Senait Gebremichael 63
Todorovska, Ana 76
Trajanov, Dimitar 51, 76
Trajkovik, Vladimir 93
Traxler, John 3
Trivodaliev, Kire 175
Tudzarski, Stojancho 51

Vodenska, Irena 76
Vrangalovski, Martin 121

Zdravkova, Katerina 135

Printed in the United States
by Baker & Taylor Publisher Services

Printed in the United States
by Baker & Taylor Publisher Services